Innovative Renewable Energy

Series editor

Ali Sayigh
World Renewable Energy Congress, Brighton, UK

The primary objective of this book series is to highlight the best-implemented worldwide policies, projects, and research dealing with renewable energy and the environment. The books are developed in partnership with the World Renewable Energy Network (WREN). WREN is one of the most effective organizations in supporting and enhancing the utilisation and implementation of renewable energy sources that are both environmentally safe and economically sustainable. Contributors to books in this series come from a worldwide network of agencies, laboratories, institutions, companies, and individuals, all working together towards an international diffusion of renewable energy technologies and applications. With contributions from most countries in the world, books in this series promote the communication and technical education of scientists, engineers, technicians and managers in this field and address the energy needs of both developing and developed countries.

Each book in the series contains contributions from WREN members and cover the most-up-to-date research developments, government policies, business models, best practices, and innovations from countries all over the globe. Additionally, the series publishes a collection of best papers presented during the annual and bi-annual World Renewable Energy Congress and Forum each year.

More information about this series at http://www.springer.com/series/15925

Ali Sayigh
Editor

The Importance of Wood and Timber in Sustainable Buildings

Springer

Editor
Ali Sayigh
World Renewable Energy Congress
Brighton, UK

ISSN 2522-8927 ISSN 2522-8935 (electronic)
Innovative Renewable Energy
ISBN 978-3-030-71699-8 ISBN 978-3-030-71700-1 (eBook)
https://doi.org/10.1007/978-3-030-71700-1

This Springer imprint is published by the registered company Springer Nature Switzerland AG
The registered company address is: Gewerbestrasse 11, 6330 Cham, Switzerland

Introduction

Faced as we are with the dire implications of climate change, there is an urgent need to discover every possible route to bring about the necessary level of reduction of carbon emissions by 2030. In this regard enormous amounts of research and development have been focused on the use of renewable energies, the replacement of fossil fuels, the control of deforestation, and finding a solution to the destruction of the environment, both land and marine, by plastic pollution. However, in view of the fact that 40%, in developed countries, and 55%, in the rest of the world, of carbon emissions are derived from the building/construction industry as a whole, with concrete alone contributing 4–8% of global carbon emissions, it is imperative that we consider all means to bring about a reduction in the levels of carbon emission. The major source of this carbon emission comes from the use of materials that have a high level of embedded carbon, such as steel, cement, and concrete. Timber and wood, on the other hand, absorb carbon and have many other beneficial attributes.

Traditionally, wood has been used in many ways in the building industry ranging from major constructional use such as wall and roofing struts to door knobs. Each use requires specific types of wood, whether it be for load-bearing or ornamentation and interiors. The important consideration nowadays is that the wood must come from a sustainable source, i.e., from forests managed to ensure continual production of new wood. Of course, it is important that timber-producing countries adhere to sustainable forestry regulations and meet the requirements of importing countries that operate sustainable forestry certification. For example, Denmark has a Program for the Endorsement of Forest Certification. More needs to be done to increase general public awareness of sustainable wood production. An increasingly valuable source of sustainable timber is the use of reclaimed timber from old buildings and disused railway sleepers, and the general recycling of wood when buildings are pulled down or refurbished. Every type of wood has its own properties with regard to hardness, shrinkage, flexibility, odor, and weight and it is for architects, builders, and designers to decide on the wood most appropriate for their project (*Wood Houses*, Francesc Zamora Mola ed, Koneman, 2018). As a building material wood has many advantages in addition to sustainability. It can be made ready for construction as a kit that is easily assembled on-site; it is easy to transport and can be cut with precision; it is a much easier material to obtain and use than steel, cement, and concrete; contrary to popular belief, thick wood is slow-burning and chars rather than combusts, whereas steel, when subject to extreme heat, can collapse. The footprint of some of the building materials, [Ref] show the following: for rolled steel it is 3778 g/kg, for copper it is 973 g/kg, for Aerated Concrete it is 1010 g/kg, for float glass 1230 g/kg, wood it is 300 g/kg. [*Carbon footprint for building products*, VTT Technology 115, edited by Antti Ruuska].

Traditional buildings in every climate zone and every culture have utilized wood. It has been the major building material alongside stone and adobe for time immemorial. With the onset of the Industrial Revolution in the late eighteenth century, wood and brick have been gradually replaced by manufactured materials such as steel, cement, and concrete. Their greatest advantage over timber and wood was thought to be because they enabled the construction of multistory buildings and eventually gave rise to high-rise buildings.

But the tragedy of 9/11 in New York and many earthquakes have made the construction industry question the invulnerability of steel and glass structures. Architects and engineers are once again looking at wooden structures.

This book emphasizes the important message that architects and structural engineers must ensure that the buildings they design and construct should not be major contributors to climate change rather they should use appropriate materials and building methods to safeguard the environment. It is of paramount importance that in the next decade carbon emissions be reduced so that the global temperature should not rise by more than 1 °C.

Contributors to this book come from 13 different countries and represent varied cultures and climate zones.

Ali Sayigh—UK 2021

Contents

About the Authors

Fadzidah Abdullah has been working in IIUM for more than 22 years. In 1998, she was among those pioneers who worked hard into establishing the Kulliyyah of Architecture and Environmental Design (KAED), International Islamic University Malaysia. Prior to becoming an academician, Dr. Fadzidah had experience working in private architectural firm for 3 years and later in Public Works Department of Malaysia (JKR) for another 3 years. Dr. Fadzidah has held some administrative posts such as Head of Department of Architecture and Academic Advisor, for both Bachelor of Architecture and Bachelor of Science in Architectural Studies.

Dr. Fadzidah Abdullah completed her Ph.D. in Architecture in 2006 at the University of Strathclyde, United Kingdom. In 2001, she was conferred with M.Sc. in Computer Aided Environmental Design (CAED) at the University of Sheffield, UK. She obtained her first degree in Bachelor of Architecture from Texas Technology University, USA, in December 1992.

Dr. Fadzidah is actively involved in doing research in the area of architectural education, architectural heritage, inclusive design, and computer-aided design (CAD). She also has interests in advancing architectural education using contemporary technology by expanding the usage of virtual reality into exploring the potential of augmented reality in CAD. Dr. Fadzidah Abdullah has published more than 100 articles in journals, conference papers, and books. She has been appointed as co-editors for several journals at national and international levels. Dr. Fadzidah has also worked collaboratively with academics in the University of Bung Hatta, Indonesia; Fatih Sultan Mehmet Waqf University, Istanbul, Turkey; and Asia-Pacific University of Bangladesh, for research on Heritage Architecture. She also involved in IIUM promotion and was sent to several countries for such purposes.

In term of architecture consultancy, Dr. Fadzidah was commissioned to do the following projects: Master Plan of Matriculation Campus in Janda baik (IIUM, 2001), Access Audit of Malaysian Main Cities (KUDU, 2008), and Master Plan of IIUM Campus (IIUM, 2013). During the last 3 years, Dr. Abdullah published more than 13 research papers.

Mohsen Aboulnaga, Ph.D. has more than 35 years of vast experience in higher education, high-level government positions, senior management, and consultancy in sustainable cities, climate change adaptation, and sustainable development as well as sustainable energy (energy efficiency and renewable energy). He is a qualified architect and graduated from Cairo University with B.Sc. Arch. (1979) and M.Sc. Housing Economic (1985). He also holds a Ph.D. degree from the University of Leeds, UK (1991) (https://www.leeds.ac.uk/). Dr. Aboulnaga is a Professor of Sustainable Built Environments at Cairo University (https://scholar.cu.edu.eg/mohsen_aboulnaga/). He is a senior expert in sustainable urban development, including green and smart cities and green building. He was involved in developing many concept notes for European Union (EU) funded projects and the Union for Mediterranean (UfM). Dr. Aboulnaga's areas of competency encompass sustainable development attainment, strategy planning and policy development, low-carbon society, and sustainable cities as well as climate change mitigation and urban farming in cities. Other areas of expertise include executive capacity building for local governments and coaching training on strategic planning, leadership, strategic planning, and SDGs and innovation.

Prof. Aboulnaga has many milestone jobs and publications, and he is WREN Pioneer and main advisor. He has more than 100 published articles in recognized journals, and he was involved in writing more than 50 books. University of Leeds, Leeds, UK

Sustainable Built Environments, Faculty of Engineering, Cairo University, Giza, Egypt

World Renewable Energy Network—WREN, Brighton, UK

Falah Al-Kubaisy, B.Arch., M.A., Ph.D. is a qualified Architect, Planner, and Urban Designer with over 40 years of experience working in several countries in the Middle East region and lecturing architectural as well as urban planning courses at universities. A winner of numerous awards, Dr. AlKubaisy has authored several published books in both Arabic and English—out of which five have been published at Amazon.com and other globally known websites.

A former Development Advisor at the Ministry of Works, Municipalities Affairs and Urban Planning in Bahrain for more than 13 years, Dr. Falah participates in managing and supervising several projects in Bahrain and the Middle East countries. He is continuously involved in various programs, assignments, and projects in strategy, policy, urban development, green building, and setting development plans for enhancing the efficiency of governmental entities in a number of countries in regions such as Bahrain, Iraq, United Arab Emirates, and Yemen.

Khaled A. Al-Sallal is a specialist in sustainable design and currently works as a Project Consultant at Concreto, a construction firm in Vancouver, Canada.

As an academic, I have more than 30 years of experience in educational and research work and have been a professor of architecture since 2012. I have published more than 100 research papers in high-impact journals and books and have given numerous talks at prestigious conferences and scientific forums such as WREC, IBPSA, ASES, and more.

My research area focuses on the interface between construction technology and design of ecologically responsible environments. I established, and served as technical manager of, the state-of-the-art Daylighting Simulation Laboratory at UAEU. I have assumed major roles in supervising Ph.D. and M.Sc. dissertations, in addition to implementing research and consulting projects that introduced new sustainable solutions.

I am an active member of IBPSA and founder and president of IBPSA-UAE, an active member of the WREN/WREC steering committee, a member of CaGBC, a full member of ASHRAE, and have served on the Board of Governors and as the Vice President, CTC Chair, and RP Chair of the ASHRAE Falcon Chapter. I have a B.Arch. from Ain Shams University, an M.S. from Arizona State University, and a Ph.D. from Texas A&M University.

Mona Azarbayjani is an Associate Professor and Graduate Program Director at the University of North Carolina at Charlotte, USA.

She obtained her buckler degree in 2003, master's degree in 2006, and Ph.D. in 2010, all from the University of Illinois at Urbana.

B.1. Academic:

Graduate Program Director 2020–

Associate Professor, Architecture, UNC-Charlotte May 2016–Present

Assistant Professor, Architecture, UNC-Charlotte 2010–2016

Graduate Research Assistant and Teaching Assistant, University of Illinois at Urbana-Champaign, 2006–2010

B.2. Non-academic:

Design Architect—Naghshe Jahan Pars Engineering Consultant Company 2003–2006

Prof. Azarbayjani has many publications in recognized journals in addition to many written chapters in books and technical reports about buildings.

Synergistic Activities

1. Research Faculty Associate, Energy Production and Infrastructure Center (EPIC), UNCC 2012–Present
2. LEED (Leadership in Energy and Environmental Design) A.P., 2009–present
3. Reviewer—Conference and Journals—BTES, ARCC conferences, *Journal of Sustainability*, *Perkins + Will research Journal*, *Energy and Buildings* journal
4. Member ASHRAE, American Society of Heating, Refrigerating and Air-Conditioning Engineers, Member SBSE, Society of Building

Science Educators 2007–present, Member IBPSA (International Building Performance Simulation Association) 2007–present

Funded Research

Solar Decathlon 2013, Design an energy-efficient solar-powered house in the competition of Solar Decathlon 2013. PIs: M. Azarbayjani, V. Cecchi, et al. Funded by DoE. Amount Raised: $1.2 Million, 2 years, 2011–2013

2012–2019: Sustainably Integrated Buildings and Sites (SIBS)—NSF Industry & University Collaborative Research Center (I/UCRC), Center Institutions: UNC Charlotte, Carnegie Mellon University, and City College of New York

Role: researcher

SIBS has approximately $400,000 in annual funding.

Rahman Azari Research Interests: Quantification of environmental impacts of built environments. Operational and embodied energy and carbon efficiency. Building envelope design, optimization for energy and environmental performance. User behavior and energy consumption

Academic Appointments: Assistant Professor, 2017—Present College of Architecture, Illinois Institute of Technology (IIT), Chicago, IL. Assistant Professor, 2013–2017 Department of Architecture, University of Texas San Antonio, San Antonio, TX. Teaching/Research Associate, 2008–2013 College of Built Environments, University of Washington in Seattle WA. Lecturer, 2007–2008 Department of Architecture, University of Kashan, IRAN. Lecturer, 2005–2007 Department of Architecture, I A. University of Tabriz, IRAN

Education: Ph.D. in Built Environment (Sustainability), 2008–2013 University of Washington, Seattle, WA Bachelor and Master of Architecture, 1996–2002 Sahand University of Technology, IRAN. He has several awards and recognition and published more than 30 papers and articles in reputable journals.

Aida Kesuma Azmin possesses B.A. Honors in Architecture [International] (UK); M.Phil. in Architecture (UK), and Ph.D. in Architecture (UK). Currently, she is an Assistant Professor at the *Kulliyyah* of Architecture and Environmental Design (KAED), International Islamic University Malaysia (IIUM). Dr. Azmin has had several architectural design practical trainings in between her A-levels and early studies in architecture semester breaks. After obtaining her Bachelor's (Hons) Degree in Architecture [International] focusing on the Northern African and Middle-Eastern Muslim countries, from the University of Huddersfield, West Yorkshire, UK, in 1998, she had worked as a design architect before turning her career as an academician. Until now, she has 20 years of teaching and research background in architecture. Dr. Azmin obtained her M.Phil. in Architecture (2002) and Ph.D. in Architecture (with Distinction) award from Heriot-Watt University, Edinburgh, in 2007. Her expertise is in Cultural Heritage Architecture—Traditional Malay Houses Architecture and Cultural Meanings. Her other research interests include Islamic Architecture, Experiential Architecture, Sustainable Design and Digital Architecture, such as Big Data, Augmented Reality, and Blended Learning Applications in Architecture. She has supervised both undergrads and postgraduate students in architecture (Part 1 and Part 2) as well as PG (Masters and Ph.D.) by research students; taught various subjects in architecture from Building Construction and Materials, Working Drawings and Design Studios to Ecological and Sustainable Design. Her main priority teaching subjects include Theory of Architecture, and History and Theory of Southeast Asia and Asian countries. Administratively, Dr. Azmin has held the Academic Advisor position between 2008 and 2018. She has gone through her postdoctoral study on Finnish woman emancipator architect, Wivi Lonn's Eco-Vernacular houses, at Jyvakylla University, Finland, in 2012. She has published 2 patented books on Architectural Heritage Studies, more than 15 books, several journal papers, articles, and many conference proceedings since 2008.

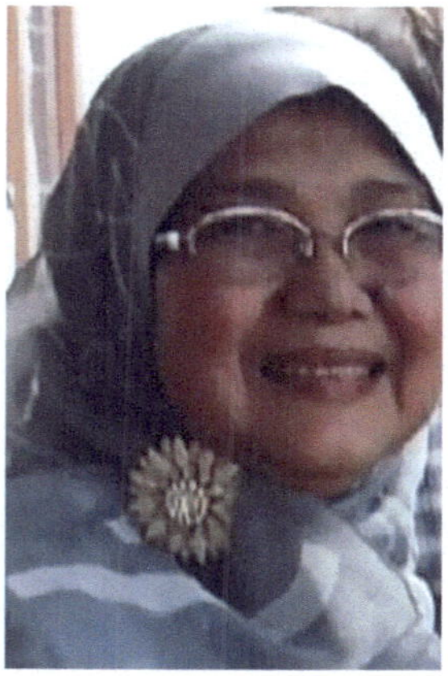

Puan Sri Datin Seri Nila Inangda Manyam Keumala Binti Haji Daud is an Expert/Research Consultant in the Department of Architecture, Faculty of Built Environment at the University of Malaya. She was one of four pioneers of Architecture Program in University Malaya, the pioneer University in Malaysia. She leads the Research Center for Urban Study, Conservation and Tropical Architecture, under Sustainability Science Cluster, University Malaya. She obtained her degrees in Bachelor of Architecture and Master of Philosophy in Architecture from University Technology of Malaysia (UTM). She was an Associate Professor and Head of the Department of Architecture before she led the Equatorial and Sustainable Design in Faculty of Built Environment, University Malaya. She is a member of the Malaysian Board of Architects (LAM), Malaysian Institute of Architects (PAM), an individual member of CIB (International Council for Research and Innovation in Building and Construction), and BRE (Building Research Establishment). She is the Advisor to the Women Graduate Association of Malaysia (PSWM) and a life member of Malaysian Heritage Trust.

Currently, she chairs the Endowment Fund Committee for the Faculty of Built Environment, University Malaya. Her research projects, supervision, and programs cover Sustainable Architectural Design, POE, Building Conservation, and Architecture of the Muslim World..

Maryam Elsharkawy, Ph.D, M.Sc., B.Sc. is a Ph.D. Candidate at Cairo University, Faculty of Engineering, Department of Architecture, with a thesis defense approval "Enhancing Daylight Utilization in Adaptive Reuse of Heritage Buildings Using Building Information Modeling." Eng. Maryam is also working a part-time Lecturer Assistant and Research Assistant at the Faculty of Engineering, Cairo University, where she is a Teaching Assistant for a course on Smart Building for the 4th year Senior 1 students. She also works at the School of Science and Engineering, American University in Cairo—AUC. She acquired a Master of Science Degree in Environmental Design from Cairo University in 2015—Thesis title: "Design Guide for Educational Building Retrofitted Facades to

Achieve Low Energy Performance in Egypt." She also graduated with a Bachelor Degree (B.Sc.) in Architecture and Engineering Technology (AET) from Cairo University in 2012. During this period she had attended two international outreach summer programs that are related to her study. Eng. Maryam worked as an Architect for ECG Engineering Consultants Group S.A. between 2012 and 2016. She received LEED Certificate from the US Green Building Council (US-GBC) in 2014. This certificate allows her to expand her professional knowledge and experience further in the field. Her experience in building performance simulation and enhancement has enabled her to carry out many research projects and professional works. She has attended many professional workshops, notably the Technical Building Construction workshop in Berlin for the application of building technologies in 2015 and similar workshops in London between 2016 and 2019. She is contributing, with Prof. Dr. Mohsen Aboulnaga, to World Renewable Energy Congress (WREC 2020) due in Lisbon by an accepted paper abstract entitled: "Integrating Building Information Modeling Tool for Energy and Daylighting Enhancement in Buildings' Reuse and Attaining Sustainability." Currently, Eng. Maryam is a member of CETL Research Team at Cairo University. Faculty of Engineering, Cairo University, Giza, Egypt

João Gomes Ferreira is an Associate Professor at the Department of Civil Engineering, Architecture and Georesources of Instituto Supeior Técnico, University of Lisbon. His main research fields deal with the rehabilitation of old buildings, especially regarding seismic strengthening, timber structures, and new structural materials. He has authored 7 books and book chapters, more than 50 papers in international journals, more than 60 communications in international conferences, and more than 350 consultancy reports. He has supervised 7 Ph.D. thesis and more than 40 Master's dissertations. He has been involved in numerous national and international research projects and presently leads two research projects in the area of strengthening old buildings and seismic protection. Dept. Ciências e Engenharia do Ambiente, Faculdade de Ciências e Tecnologia, Quinta da Torre, Monte de Caparica, Portugal

Manuel Correia Guedes is director of the Architectural Research Centre of the Instituto Superior Tecnico (ICIST—Group 8). He is responsible for several disciplines of the courses of Architecture, Civil Engineering, and Territorial Engineering. He is supervising several Ph.D. and M.Sc. students.

He was Chief Coordinator of a COOPENER E.U. project (SUREAFRICA), National Coordinator of an ASIA-LINK project, and has participated in various international and national conferences, seminars, and workshops.

His Ph.D. was on Environmental Design in Southern European Office Buildings covering sustainability, passive design, energy efficiency, and thermal comfort. Both his Ph.D. and M.Phil. were done at Cambridge University, Martin Centre. His M.Phil. thesis was sponsored by the E.U. "Praxis" scholarship on Environmental Design, "Aspects of Environmental Design of Office Buildings in Portugal."

Since 1985 he participated in various projects, namely the Portuguese Pavilion in Seville's EXPO 92, two residential buildings in Vila Real, the competition for the National Assembly building, and the building of the Agronomy Faculty (UTL). He worked as an architect in several Portuguese architectural companies. He published many articles and papers in the fields of bioclimatic architecture and the built environment. Department of Civil Engineering and Architecture, Instituto Superior Tecnico, Lisbon, Portugal

Architectural Research Centre (ICIST-N8), Lisbon, Portugal

Carolina Ganem Karlen is a Senior Lecturer in the School of Design at the National University of Cuyo. (FAD-UNCuyo), Mendoza, Argentina, and a Senior Researcher at the National Research Council of Argentina (INAHE-CONICET). She is also the Academic Director of the Master in Sustainable Architecture at the Congress University (UC). Prof. Ganem Karlen studied architecture at the University of Mendoza (FAU-UM) where she obtained her degree in 2001. That same year, she started her postgraduate studies in ABITA Centre at the Universitd degli Studi di Firenze (UNIFI), Florence, Italy with Professor Marco Sala. She holds a Ph.D. (2006) from the Universitat Polit6cnica de Catalunya (ETSAB-UPC), Barcelona, Spain, under the guidance of the late Professor Rafael Serra Florenza and Professor Helena Coch Roura. Prof. Ganem Karlen's research focuses are on energy efficiency, indoor environment and well-being, micro-urban environments, occupants' thermal comfort and adaptive behavior, and energy assessment and certification of buildings. She has completed numerous research projects using a variety of resources. Prof. Ganem Karlen has authored or coauthored many book chapters and more than 150 papers and research reports mainly presented in international journals or at international conferences with referee. Prof. Ganem Karlen has been a member of the scientific committee and organization committee for many international conferences. Referee for many international journals and international conferences in the field of energy and building. Prof. Ganem Karlen is a Chartered Architect, a Fellow of the International Society of Building Science Educators (SBSE), a Fellow of the International Building Performance Simulation Association Latin-American Chapter (IBPSA-Latam), and a Fellow of the Argentinean Association of Renewable Energy (ASADES).

Luis Morgado is a Portuguese Architect based in Toronto, Canada. He worked as Architectural Designer with "architectsAlliance," "Quadrangle," and "Monteyne Architecture Works." In Portugal he owned the studio ALM arquitectos associados. In 2016 he obtained the Doctorate Degree in Architecture, at Instituto Superior Técnico, with the thesis: "Wood houses in Portugal—Architectural design methodology." He was part of the research team for the study "Housing for the Future," at Laboratório Nacional de Engenharia Civil, with the study "Emergent Housing Types." From 1999 to 2011 he was a lecturer at Universidade Católica Portuguesa, Instituto Superior Manuel Teixeira Gomes, and Fundação Ricardo Espírito Santo. Travessa do Carvalho 23, 1249-003 Lisboa, Portugal.

Muhammad Faizal Bin Abdul Rani obtained his Ph.D. **in** Thai Studies at Chulalongkorn University in 2018, Bangkok, Thailand, his M.Sc. in Building Technology and Construction from the University of Malaya in 2007, Kuala Lumpur, Malaysia, and his B.Sc. (Honor) in Interior Design from the University of Science, Malaysia, in 2003, Penang, Malaysia. Presently, Dr. Rani is an experienced senior lecturer with over 13 years of experience in academic and interior design profession. He has excellent reputation for resolving problems, improving customer satisfaction, and driving overall operational improvements and has consistently saved costs while increasing profits. Dr. Rani published many well-known papers in various journals and won several awards for his innovative architectural design and teaching status. He is a member of Malaysian Institute of Interior Designers, Graphic Design Association of Malaysia, and the Siam Society under Royal Patronage.

Seyedehmamak Salavatian, Ph.D. in architecture, faculty member of Architecture department, Islamic Azad University, Rasht Branch, Rasht, Iran, was born in 1983 and received both Bachelor's and Master's degrees in architecture in 2005 and 2008, respectively, at a high-ranking university (Shahid Beheshti) in Tehran, Iran. Her final thesis was done on the design principles of sustainable metro stations. She has also completed her Ph.D. studies in building thermal performance at universita politecnica delle Marche in 2016. Her Doctoral Dissertation was focused on the energy-efficient design of temporary prefabricated homes, which was the subject of a book published subsequently. She has a 5-year experience in engineering consultant companies as the designer of residential and industrial building projects. She has also participated in the conceptual phase of the national observatory building and preliminary and detailed phases of international airport terminal as well as a number of refinery plant buildings. In addition, a number of architectural books have been translated into Persian by her and published which are helpful for architecture students. She has research experiences in subjects such as vernacular architecture, sustainable insulation material for buildings, life cycle assessment and life cycle cost of buildings, uncertainty evaluation in building global cost, and sunshade design which involves both simulation and experimental tests. These studies were reflected in scientific papers and presented in international technical conferences. Since 2016 she started working as a faculty member of architecture department at Islamic Azad University and lectures in various courses such as architectural design and theoretical courses as bioclimatic architecture, research methods, and construction materials. Hence, numerous Master's theses as well as a number of doctoral dissertations were carried out under her supervision and consultancy.

Aliyah Nur Zafirah Sanusi, B.Sc., B.Arch., M.Phil. is currently an Assistant Professor at the Kulliyyah of Architecture and Environmental Design (KAED), International Islamic University Malaysia. Aliyah Nur Zafirah Sanusi has an academic background in architecture, having obtained her B.Sc. in Architectural Studies and Bachelor of Architecture from Cardiff University (UK), which are equivalent to RIBA Part 1 and Part 2. She then furthered her studies and obtained M.Phil. in Built Environment from the University of Nottingham (UK) and then a Ph.D. degree in Energy and Sustainable Development from De Montfort University (UK).

Dr. Sanusi published more than 12 articles in recognized journals, presented more than 20 papers at conferences, and written chapters in more than 8 books.

Ali Sayigh is a UK Citizen and graduated from London University & Imperial College, BSC.AWP, DIC, Ph.D., in 1966. He is Fellow of the Institute of Energy and Fellow of the Institution of Engineering & Technology, Chartered Engineer, Chairman of Iraq Energy Institute, and Fellow of the Royal Society of Arts.

Prof. Sayigh taught at various institutes in Iraq, Saudi Arabia, and Kuwait as well as in Reading University and University of Hertfordshire from 1966 to 2004. He was Head of Energy Department at Kuwait Institute for Scientific Research (KISR) and Expert in renewable energy at AOPEC, Kuwait, from 1981 to 1985.

He started working in solar energy in September 1969. In 1972, he established with some colleagues in Saudi Arabia *The Journal of Engineering Sciences* in Riyadh, Saudi Arabia, and in 1984 he established *International Journal for Solar and Wind Technology*, as an editor-in-chief. This has been renamed in 1990 to *Journal of Renewable Energy*. He is the editor of several international journals published in Morocco, Iran, Bangladesh, Nigeria, and India. He established WREN and the World Renewable Energy Congress in 1990. He is a member of various societies related to climate change and renewable energy. He is Chairman of Iraq Energy Institute since 2010.

He was a consultant to many national and international organizations, among them, the British Council, ISESCO, UNESCO, UNDP, ESCWA, UNIDO, and UN. He organized conferences and seminars in 54 different countries and published more than 600 papers. He edited, written, and is associated with more than 100 books. He supervised more than 82 M.Sc. and 36 Ph.D. students. He is the editor-in-chief of the yearly *Renewable Energy Magazine*, 2000–2016. He is the founder of *WREN* and *Renewable Energy Journal* published by Elsevier and was the editor-in-chief for 30 years from 1984 to 2014.

He is the editor-in-chief of *Comprehensive Renewable Energy* coordinating 154 top scientists, engineers, and researchers' contribution in 8 volumes published in 2012 by Elsevier which won the 2013 PROSE award in the USA. He is the founder of Med Green Buildings and Renewable Energy Forum since 2011. In 2016 he established a peer-reviewed international open access journal called *Renewable Energy and Environmental Sustainability*—REES, which is published in English online by EDP publisher in Paris.

He was the winner of the Best Clean Energy Implementation Support NPO—UK. In 2018 WREN was rated globally as one of the best organizations in the UK promoting renewable energy. In November 2018, Prof. Sayigh was elected fellow of the Royal Society of Art (FRSA). Prof. Sayigh works with Springer Nature in publishing books and proceedings since 2014 and up to now.

Maryam Singery has over 17 years of experience, and she holds a Ph.D. degree and teaches at South Texas College as a full-time lecturer. Her areas of specialization and teaching focus on sustainable architecture and urban studies both internationally and in different universities. She has published articles, book chapters, a book (*Urban Sustainable Environment*) and has presented in numerous international conferences. She has also been an active member of several environmental groups in Iran and has been the recipient of several awards and honors in her fields of expertise. She extracts and incorporates valuable sources of inspiration from sustainable architecture and urbanism into her teaching.

David Jacob Thaddeus Education

1985–1988 Master of Architecture, College of Architecture, the University of Houston

1977–1981 Bachelor of Civil Engineering, the American University of Beirut, Lebanon

Academic Appointments

Teaching Assignments at the College of Architecture, the University of North Carolina-Charlotte

2008–Present Professor

Architectural Structures and Architectural Design

1999–2008 Associate Professor. Architectural Structures and Architectural Design Teaching Assignments at the Gerald D. Hines College of Architecture, the University of Houston

1997–1999 Associate Professor (with tenure), undergraduate and graduate courses in Architectural Structures and Architectural Design

1991–1997 Assistant Professor, undergraduate and graduate courses in Architectural Structures and Architectural Design

1989–1991 Visiting Assistant Professor, Architectural Structures courses

Additional Teaching Activities:

Conducted many seminars on the Structural Divisions of the Architect Registration Examination (A.R.E.) from 1980 to 2003, in various countries

Research, Publication, and Grants (selected)

2010—Present TOYS: Visual Teaching and Learning Using Digitally-Fabricated and 3D Printed Structural Models and Animations

2008—Present Visual Glossary and Mock Exam questions for the Architect Registration Exam (ARE) (www.davidthaddeus.com)

2004–2006 Development of a Curriculum for the Instruction of Structural Steel Topics in Colleges of Architecture

External Funding from the American Institute of Steel Construction-AISC

2004 Composed new cards and edited existing cards of the General Structures and Lateral Forces Components of the Archi*flash* card study system for the Architect Registration Exam (ARE). Archif*lash* System. NALSA Publishing 2005

2001 Visualizing the Structural Behavior of Buildings and Long-Span Structures. Design and construction of a wind tunnel for illustrating wind forces and a shake table for visualizing seismic forces

1993–1995 Spreadsheets for Architects, coauthored with Leonard Bachman, University of Houston Book and accompanying software, Van Nostrand Reinhold, 1995 NCARB, Washington, DC, USA

Antonella Trombadore is qualified as Associate Professor and is a researcher at the University of Florence. Since 1999 she works at the Department of Architecture and ABITA Interuniversity research center in the field of Sustainable Architecture, Responsive Design, Green Architecture for Resilient Cities, and Sustainable Immersive tourism. She is team leader of several European Research Projects in the field of architectural integration of innovative solutions for energy retrofitting action, nearly zero buildings as well as integrated smart process in Mediterranean climatic and cultural context.

Technical coordinator of international Post-Graduated Master Course in "Sustainable Architecture," she is especially involved in the design and technologies for built environment in the different Mediterranean and North Africa climatic zone, with a focus on nature-based solutions for climate and environmental responsive design. She is contract expert for Research Executive Agency (REA) European Commission (2017) for the evaluation of international research project HORIZON–EeB-05-2017—Development of Near

Zero Energy Building Renovation; member of international network as PLEA and WREC, and author of scientific contributions, papers, and books.

Zeenat Begam Yusof is a Senior Lecturer at the International Islamic University (IIUM) Malaysia under the School of Architecture and Environmental Design. She is also the Academic Advisor of Architecture Department. Dr. Yusof studied architecture at the IIUM where she obtained her B.Arch. degree in 2004. After 3 years, she started her Master of Environment study at Universiti Putra Malaysia. Then she finished her Ph.D. in 2017 from the Universiti Teknologi Mara Malaysia. Her research focuses on green operation practices for hospitality industry, sustainable tourism, marine tourism, environmental impact studies on marine environment, and sustainable building materials.

She has completed several research projects with the help of a wide variety of grants from Ministry and University. Dr. Yusof has authored and coauthored several journal articles and book chapters which are more than 20 after her Ph.D. studies. She has presented her research at international conferences such as in Istanbul, London, Barcelona, and Hanoi Vietnam.

Dr. Yusof won five awards at the national level and two awards at the international level in her field. She has also been a member of the scientific committee and organizational committee for several conferences. She is a referee for few international journals in the field of sustainable tourism, one of which is *Tourism Review*. Dr. Yusof is a Graduate Architect and Technologist in the area of building construction.

Chapter 1
Bamboo: The Forgotten Versatile Materials

Fadzidah Abdullah, Aliyah Nur Zafirah Sanusi, Aida Kesuma Azmin, and Zeenat Begam Yusof

1.1 Introduction

In the modern construction industry, there is an emergence of many new construction materials that come with considerably high monetary value. This phenomenon contributes to the increase of building construction cost, yet most modern buildings in tropical regions still require excessive energy usage to provide comfort, usually by means of installing air-conditioned systems to cool the interior spaces. Hence, there is a question of why the construction industry does not turn to traditional materials like bamboo, which are known to provide better thermal comfort for buildings' occupants and are considered as environment-friendly materials as well. Perhaps, it is time for the building construction industry to make a big U-turn, considering ways of using bamboo in modern building construction for the benefit of energy saving and sustainability of heritage materials.

As a forgotten versatile building material, bamboo deserves to be reestablished in the construction industry, for the construction of both modern and traditional architecture. Thence, this chapter intends to enquire about the potentials of bamboo as building materials and to evaluate the strategies of using bamboo based on case studies of existing bamboo buildings throughout the tropical regions of Southeast Asia.

In most Asian countries, bamboo has been used for many aspects of life, ranging from household products to building structure. Even though it has varieties of usage, bamboo is known as the "poor man's timber" [1] not only in India but also in most Southeast Asian countries like Malaysia, Indonesia, Thailand, Myanmar, Vietnam, and the Philippines. Nevertheless, Far East countries like China, Korea, and Japan

F. Abdullah (✉) · A. N. Z. Sanusi · A. K. Azmin · Z. B. Yusof
International Islamic University Malaysia, Jalan Gombak, Kuala Lumpur, Malaysia
e-mail: fadzidah@iium.edu.my

A. Sayigh (ed.), *The Importance of Wood and Timber in Sustainable Buildings*,
Innovative Renewable Energy, https://doi.org/10.1007/978-3-030-71700-1_

have used bamboo with remarkable versatility that the people regard bamboo as part of the representation of their culture. In Japan, bamboo is a symbol of prosperity because of its sturdy root structure [2]. Bamboo also symbolizes purity and innocence, where the Japanese associate bamboo with a proverb, "Take o watta youna hito," which figuratively means a man with a frank nature is represented by fresh-split bamboo [2].

The Japanese have inattentiveness toward bamboo for its flexibility and strength, and the usage of bamboo is apparent in both their traditional and modern architecture. Nonetheless, it is the Chinese who have revolutionized the utilization of bamboo in their design creativity, not only for the construction industry but also for the creative production of sunglasses and bras. This "bamboo revolution" has spawned the idea of bamboo as the "timber of the future" [3].

With the global increasing threat of deforestation as demand for wood rises, Lancaster [3] perceives that bamboo could be touted as a high-potential substitute for timber in the construction industry. In Southeast Asian countries, there are about 500 species of bamboo that could cater the increasing need of raw bamboo, especially for the construction industry. Hence, Southeast Asia has the prime position to capitalize the commercialism of this neo-timber, for the increasing demands for bamboo products throughout the globe [3].

1.2 Bamboo as Building Materials

The usage of bamboos as building materials and in construction has many advantages and some disadvantages. However, the benefits overcome the drawbacks, and it is one of the sustainable materials for buildings. Bamboo has a fast growth rate, which is faster than most of the other plants [4].

Advantages

The advantage of using bamboos in building structure and construction is that they have higher tensile strengths as compared to steel because of their fibers that run axially. Furthermore, bamboos are considered as lightweight materials which make them easier to transport, construct, and install. Their flexible and elastic nature is useful for buildings in earthquake-prone areas. Their elasticity is also an advantage that enables architects to design organic form buildings and easily construct with the bent bamboo materials. Apart from its strength, lightweight, and elasticity, bamboos are able to resist fire up to 4000 °C. This is due to their contents, which has a high value of silicate acid and water. However, these materials are safe to human health and do not release toxic fumes to the environment [4].

In the construction industry, bamboo is considered a suitable substitute for wood because of its superior regeneration ability, its good properties for wide use, and its short rotation period [9]. These advantages are substantiated by the fact that bamboo could be easily found all over Southeast Asia, making its cultivation sustainable.

Disadvantages

Apart from their advantages, there are also some disadvantages in using bamboos as building materials. As natural plants, bamboo's durability lasts only for 2 years naturally and up to 7 years if placed under cover [5]. Therefore, to increase their durability for a longer time, they need to be preserved before being used for building construction. Untreated, the bamboos tend to shrink when they lose their water content. Furthermore, untreated bamboos also contain starch which could attract termites, mold, and fungi. Another drawback in bamboo construction is in their jointing. The joints have to be carefully crafted as not to crack the whole bamboo [6].

Bamboo Treatment and Preservation

Despite the disadvantages of bamboo usage as building material, recent technology has proven that it is possible to lengthen bamboo lifespan and strengthen its durability. Bamboo could be soaked in a pool of boric acid solution for a week to allow the solution to fully penetrate inside and then dried vertically under the sun to turn the bamboo into a beautiful yellowish color. This process not only increases the durability of bamboo but also provides natural insect repellent and preservative for the bamboo [7]. Besides, the application of several types of natural oils could also help treat and preserve bamboo. Among the popular types of oil to increase bamboo durability are neem oil, cashew oil, and cedar oil.

Bamboo, as a versatile and environment-friendly material, should be treated correctly in order to increase its durability, utilization, and popularity. It is important to know that, by applying the appropriate treatments, the shelf-life of bamboo could be increased up to 50 years. Hence, the usage of bamboo is economical and sustainable [5].

Traditionally, several methods of bamboo treatment have been implemented especially by rural folks of several countries, and the systems have been passed down to bamboo growers according to traditional practices. Kaur [8] specifies that there are four main methods of traditional preservation of bamboo: good harvesting practices, water leaching method, smoke treatment, and special construction practices.

First, good harvesting practices indicates that harvesting time affects the durability of bamboo, and the right time to harvest bamboo is during or after the rainy season because the more moisture content in the bamboo meat means the less susceptible it is to fungi attack. Moisture content in bamboo reduces the starch content, simultaneously decreasing the potential of fungi attack. Second, the water leaching method involves the process of submerging the bamboo culms in running water for a duration of 3 months. This process intends to wash out starch, carbohydrates, and other water-soluble substances and eventually contributes to the enhancement of bamboo fungal resistance.

Third, smoke treatment is another method of bamboo treatment that has been used for thousands of years to enhance the durability of bamboo. This process of using smoke causes the bamboo culms to be fumigated, and thus, the bamboo becomes inedible to insects. Research has shown that bamboo culms that have been smoked for 8 h have significant reduction of starch content, hence becoming beneficial for enhancing the service life of bamboo. Lastly, special construction practices should be applied to the building constructional method, where bamboo components should be installed without direct contact with the soil and ground. This practice reduces the uptake of moisture through the soil or air. Furthermore, applying mud coating or cement plastering at the points of contact with the ground could help reduce microbial attack too.

Potential of Usages

Traditionally, bamboo has been used for building construction for thousands of years. Similarly, in modern architecture, bamboo has many potential uses in building construction, such as building structure, building skin, interior finishes, and external shading devices. However, suitable treatment must be applied beforehand in order to increase the durability of bamboo. When treated, bamboo has the same qualities as wood and could be used to provide good raw material to replace wood in manufacturing woven plywood, floorboards, particleboard, or fiberboard [9, 10]. In addition, bamboo culms can be used to produce decorative building elements, such as ceilings [9] and built-in furniture.

1.3 Bamboo in Vernacular Architecture

Traditionally, bamboo has been used as the main material for vernacular buildings in Southeast Asian regions. The usage covers a wide range of building components, from the structural system to building finishes. Unfortunately, there are not many bamboo buildings that can last for more than three decades. Today, it is quite difficult to find steady bamboo buildings of vernacular architecture especially in Malaysia. Amazingly, in Indonesia, nevertheless, some bamboo buildings have lasted for more than three centuries. The authentication of the buildings could be ascertained even until these days, and this phenomenon is proven to indicate that bamboo could actually last very long if treated correctly.

The Bayan Beleq Mosque, Lombok, Indonesia

The Bayan Beleq Mosque is one of the authentic bamboo buildings that is still in existence in Indonesia. Dated back to 1634 [11], it is considered the oldest mosque built in the island of Lombok, Indonesia. The mosque is more commonly known as *Masjid Kuno*, which means the Old Mosque. It is located in a very remote village, named *Desa Bayan*, about 80 km from the capital of Lombok, Nusa Tenggara Barat [12].

Fig. 1.1 The architectural form of the Bayan Beleq Mosque

Fig. 1.2 The architectural form of the Bayan Beleq Mosque

Built on top of a hill, the Bayan Beleq Mosque has some similar architectural features like other famous vernacular mosques in Indonesia, such as the Damak Mosque. The Bayan Beleq Mosque is surrounded by fences, with fortresses like stone walls. The mosque is designed in square shape form and has a steep top roof with two-tiered pyramidal shape. Figures 1.1 and 1.2 show the architectural form of the mosque.

Structure System

Structurally, the Bayan Beleq Mosque is constructed with stone foundation, and it has four main timber columns at the center of the building to support the upper pyramidal roof tier. There are 28 other shorter timber columns, with the height of 4.6 m, at the perimeter of the mosques, cladded with woven bamboo walls. In this building, bamboo is used as the secondary structural system. The main cornered rafters of the pyramidal shape roof are made of timber, and the secondary rafters placed in between are made of bamboo. Figure 1.3 shows a timber rafter protruded out of the roof system, and Fig. 1.4 shows the interior view of the mosque where timber and bamboo are both used for the structural system of the building. Hence, it is observed that the vernacular bamboo building relies on timber columns as the main structure, and bamboo is used for the secondary structural system for the building.

Fig. 1.3 The protruded timber rafter

Fig. 1.4 The interior view shows part of the structural system of the Bayan Beleq Mosque

Other Building Components of the Mosque

Most of the other building components used for the Bayan Beleq Mosque are made of bamboo. The wall is made of woven bamboo strips that also served as non-load-bearing walls. These four-sided walls are placed on top of the stone foundation. Figure 1.5 shows how bamboo could be woven beautifully to create a desirable pattern to decorate building facades.

In addition, the roof of the mosque has three layers of materials, creating a very thick roof. The lowest layers are bamboo substructural systems, consisting of bamboo rafters and battens. The second layer is where bamboo cuts are placed, shaped, and arranged like concave-shaped tiles. These bamboo tiles could prevent the infiltration of rainwater into the interior of the mosque. The uppermost layer is where a natural roof material named *ijuk* (coco fiber) is placed. Figure 1.6 shows the arrangement of the multiple layers of roof materials.

Fig. 1.5 Woven bamboo for the lightweight wall system

Fig. 1.6 The multilayers of bamboo roof tiles

This showcase study of a vernacular architecture has shown that bamboo is not only used for the main structural system but also for the nonstructural components of the building like the non-load-bearing wall, roof tiles, flooring, and furnishes. Hence, for a bamboo vernacular building to last long, it needs to have hardwood timber columns, beams, and rafters to support the buildings. Nevertheless, in many bamboo modern buildings, bamboo structural systems have been used innovatively by enhancing bamboo structural property at its best.

1.4 Bamboo Usage in Modern Buildings

The world has become more aware and conscious of the environmental impact caused by modern developments. Therefore, sustainable materials such as bamboo were introduced back into the construction of modern buildings in various ways: building structure, facade design, and interior floor and wall finishes.

Bamboo Sports Hall, Panyaden International School, Thailand

The Bamboo Sports Hall was designed for Panyaden International School in Chiang Mai, Thailand. The building was designed by Chiangmai Life Construction and completed in 2017 [13] (Fig. 1.7). Its spacious volume can fit a capacity of 300 students (Fig. 1.8). The vaulted hall can house various sports courts such as futsal, basketball, and badminton courts [14].

Fig. 1.7 External view of the Bamboo Sports Hall. (Source: archdaily.com 2017)

Fig. 1.8 Internal view of the Bamboo Sports Hall. (Source: archdaily.com 2017)

Structure Stability

The hall's long-span structure comprises prefabricated bamboo structure that spans 17.22 m at the height of 10.8 m (Fig. 1.9). These structures stand strong, without any steel reinforcement nor steel connectors (Fig. 1.10).

The facade design allows natural ventilation into the large space, and it is also claimed as a zero-carbon footprint during its construction because of the bamboo's natural ability to absorb carbon dioxide even as building materials [14].

Bamboo Veil House, Vietnam

The Bamboo Veil House is a modern tropical house [15] located in Singapore and designed by Wallflower Architecture and Design. It is a 680 m^2 semi-detached house completed in 2019. This house uses bamboo as part of its facade design in response to the local tropical climate (Fig. 1.11). The bamboo elements are wrapped around and create a soft facade and organic layer in between the well-defined white concrete [16].

The long thin bamboo characteristics allow the panels to be flexible and create a curved edge. Some parts of the bamboo screen are operable to allow natural ventilation flow into the indoor space (Fig. 1.12).

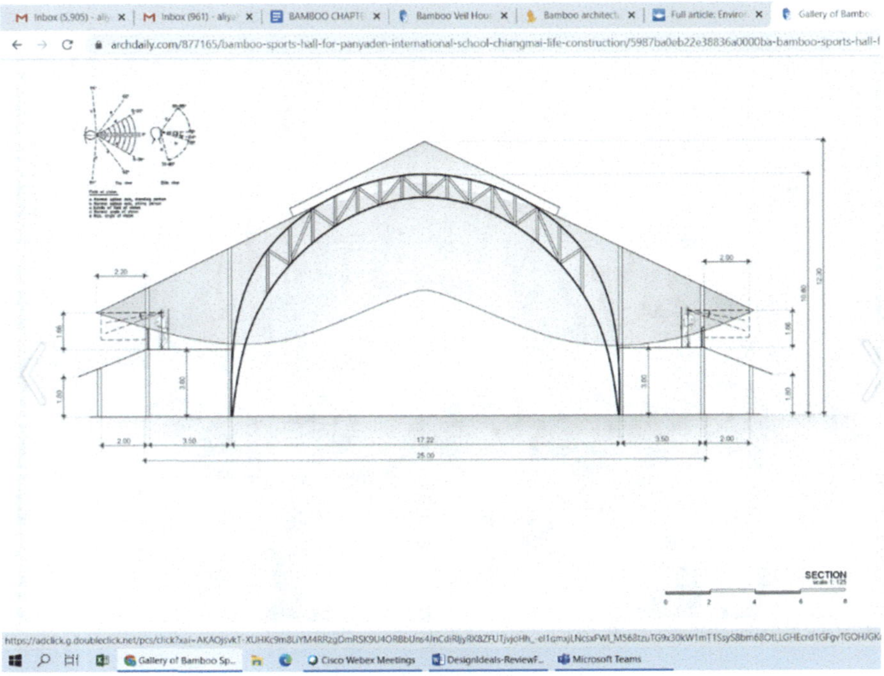

Fig. 1.9 Structural detail of the Bamboo Sports Hall. (Source: archdaily.com 2017)

Fig. 1.10 Structural joints. (Source: archdaily.com 2017)

Fig. 1.11 The external view of the Bamboo Veil House (https://www.homedit.com, accessed Sep. 09, 2020)

Fig. 1.12 The operable bamboo screens in its facade design (https://www.homedit.com, accessed Sep. 09, 2020)

Fig. 1.13 The operable bamboo panel view in the evening (Angelopoulou [16])

Furthermore, the air can also infiltrate into the indoor space through the small gaps in between the thin bamboo strips (Fig. 1.13). The usage of cladded bamboo panels for facade design can be regarded as an innovative solution of passive design strategies, for buildings in tropical contexts.

Malaysia Bamboo Pavilion, Kuala Lumpur, Malaysia

On a smaller scale, bamboo materials are also used for modern pavilion and temporary structures. One example is the Malaysia Bamboo Pavilion, designed by Ar. Dr. Eleena Jamil. Its colorful and warm bamboo features stand out from the normal concrete and steel buildings surrounding it (Fig. 1.14).

The pavilion was built in 2018; it comprises bamboo structure and bamboo finishes with colorful elements (Fig. 1.15). Cut recycled bamboos were used as its wall finishes, which is a mixture of solids and voids to allow natural ventilation through the pavilion (Fig. 1.16). The wall finish composition makes a unique colorful photo backdrop for the passersby and tourists (Fig. 1.17).

Fig. 1.14 Malaysia Bamboo pavilion, Kuala Lumpur. (Source: Reuland [17])

Fig. 1.15 Colorful bamboo wall finishes. (Source: Reuland [17])

It is a post and beam structure, where bamboos were tied together in knots. The facade was made of modular components that hold the cut recycled colorful bamboos in a frame (Fig. 1.18). This bamboo pavilion is a simple structure, yet it became an attractive feature for the urbanscapes of Kuala Lumpur.

Fig. 1.16 Solids and voids in the wall finishes. (Source: Reuland [17])

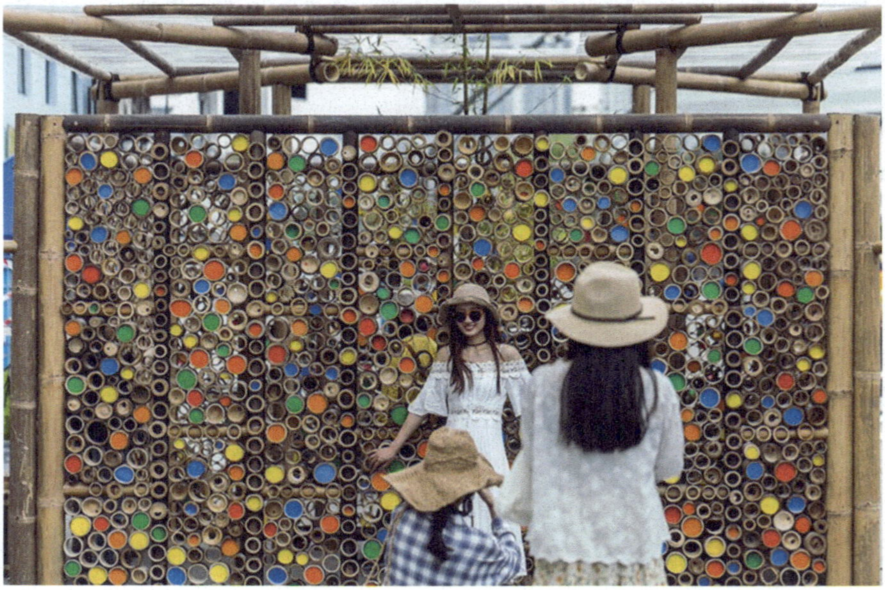

Fig. 1.17 Creating a unique and colorful photo backdrop. (Source: Reuland [17])

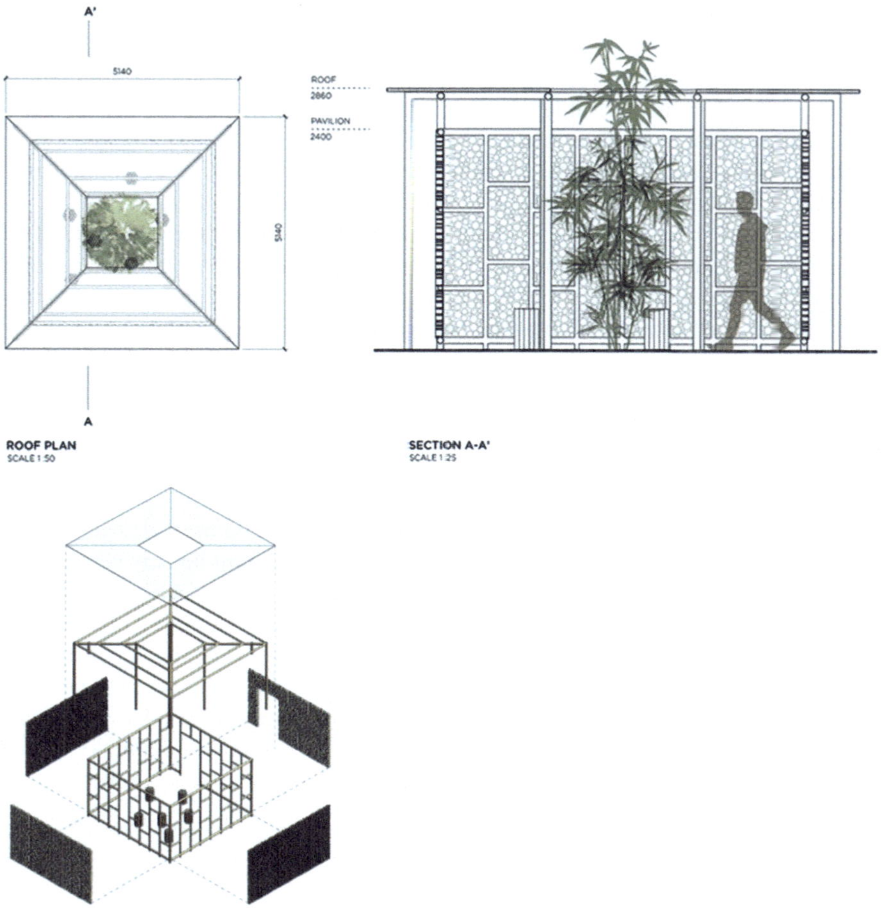

Fig. 1.18 Cross section and axonometric of the bamboo pavilion. (Source: Reuland [17])

1.5 Strategies for Bamboo Usage

The bamboo buildings discussed in this chapter have shown that bamboo can be versatilely used as the main building material. Besides enhancing the aesthetic values of architecture, bamboo could also be used using several strategies.

- Understanding the bamboo type is important to know the quality and durability of bamboo. Since there are more than a thousand types of bamboo trees, seeking advice from bamboo experts is necessary.
- Bamboo needs to be treated well by either using traditional methods or modern techniques.

- For structural systems, it is not advisable to use bamboo as standalone structural component, but combining several bamboos, tied and twisted in certain ways, could help strengthen its structural property.
- For jointings, it is advisable to use knots rather than nails or screw, because punching holes on bamboo would weaken its quality.
- Bamboo shall be mostly used for nonstructural components, as the components could be replaced without dismantling the whole building.
- Bamboo shall be used to give a positive impact to the environment because of its capability of absorbing carbon dioxide from the atmosphere. Its sequestering property traps the carbon from the atmosphere and stores it in its natural form [18]. Its carbon sequestration property contributes to reducing global warming.

1.6 Conclusion

This chapter concludes that the relevance of bamboo as building materials is very distinctive. The potential for bamboo usage as material for buildings is very broad, ranging from structural system to building finishes. The usage is suitable not only for traditional or vernacular buildings but also for modern building construction. This regeneration of bamboo usage could support the United Nation's call for the sustainable development goal (SDG) as well. Thence, bamboo deserves to be reestablished in the construction industry for the construction of both modern and traditional architecture. The attempt of reestablishing the usage of bamboo in the construction industry would also help reduce global dependency to timber supply from the global forest.

References

1. Teron, R., & Borthakur, S. K. (2012). Traditional uses of bamboo among the Karbis, a hill tribe in India. *Bamboo Science and Culture: The Journal of the American Bamboo Society, 25*(1), 1–8. © copyright 2012 by the American Bamboo Society.
2. Abe, N. (2019). Bamboo in Japanese culture. *ThoughtCo*. Retrieved September 9, 2020, from https://www.thoughtco.com/bamboo-in-japanese-culture-2028043
3. Lancaster, C. (2020). Southeast Asia is sitting on a bamboo bounty, it just doesn't know it yet. In *GLOBE-Lines of Thought Across Southeast Asia*. Retrieved September 8, 2020, from https://southeastasiaglobe.com/green-gold/
4. Jain, V. (2020). *Bamboo as a building material—Its uses and advantages in construction works*. The constructor—civil engineer home. Retrieved July 30, 2020, from https://theconstructor.org/building/bamboo-as-a-building-material-uses-advantages/14838/
5. Guadua. (2020). Natural bamboo durability. *Guadua Bamboo*. Retrieved September 9, 2020, from https://www.guaduabamboo.com/blog/durability-of-bamboo
6. Schröder, S. (2020). Durability of bamboo. *Guadua bamboo*. Retrieved July 30, 2020, from https://www.guaduabamboo.com/blog/durability-of-bamboo

7. CLALC. (2020). Bamboo treatment. CLALC—Chiangmai life architects, life construction. Retrieved September 9, 2020, from https://www.bamboo-earth-architecture-construction.com/bamboo-treatment/
8. Kaur, P. J., Satya, S., Pant, K. K., & Naik, S. N. (2016). Eco-friendly preservation of bamboo species: Traditional to modern techniques. *BioResources, 11*(4), 10604–10624.
9. Yang, Y., Wang, K., Shengji, P., & Jiming, H. (2004). Bamboo diversity and traditional uses in Yunnan, China. *Mountain Research and Development, 24*(2), 157–165.
10. Hui, C. M., and Yang, Y. M. (1998). Bamboo Timber Resources, Yunnan Science and Technology Publishing House, Kunming.
11. Wikipedia (2020). Bayan Beleq Mosque. Retrieved 22 September 2020.
12. Yulia Lisnawati (2018). Berusia 3 Abad, Masjid Kuno di Lombok Tetap Kokoh Meski Diguncang Gempa. Liputan 6. Retrieved on 16th. September 2020, from https://www.liputan6.com/citizen6/read/3617898/berusia-3-abad-masjid-kuno-di-lombok-tetap-kokoh-meski-diguncang-gempa.
13. ArchDaily. (2020). Bamboo Sports Hall for Panyaden International School/Chiangmai Life Construction (2017). *ArchDaily*. ISSN 0719–8884. Retrieved September 8, 2020, from https://www.archdaily.com/877165/bamboo-sports-hall-for-panyaden-international-school-chiangmai-life-construction
14. Xu, Y. (2018). Bamboo architecture in Southeast Asia. Retrieved September 8, 2020, from https://www.silverkris.com/bamboo-architecture/
15. Tropical house with a bamboo veil wrapped around its middle section. Retrieved September 9, 2020, from https://www.homedit.com/tropical-house-with-a-bamboo-veil/
16. Angelopoulou, S. L. (2019). Wallflower wraps house in Singapore in a bamboo veil of operable screen. *Designboom*. Retrieved September 9, 2020, from https://www.designboom.com/architecture/wallflower-house-singapore-bamboo-veil-screens-12-12-2019/
17. Reuland, R. (2018). Eleena Jamil's pavilion in Malaysia is a colourful take on bamboo. *Designmboom*. Retrieved September 9, 2020, from https://www.designboom.com/architecture/eleena-jamila-bamboo-pavilion-malaysia-03-11-2018/
18. Manandhar, R., Kim, J. H., & Kim, J. T. (2019). Environmental, social and economic sustainability of bamboo and bamboo-based construction materials in buildings. *Building Structures and Materials, 18*(2), 49–59.

Chapter 2
The Role of Wood in Current Sustainable Building in Thailand as Architectural Ornaments

Muhammad Faizal Bin Abdul Rani
and Puan Sri Datin Seri Nila Inangda Manyam Keumala Binti Haji Daud

2.1 Introduction

Background of the Topic

Thailand was known as one of the top tourist destinations in the world. Over the last 16 years (Fig. 2.1), tourist arrival in Thailand has steadily increased. However, this number has dropped since the Covid-19 pandemic has emerged in 2020.

Before the Covid-19 pandemic appeared, Thailand has successfully launched many campaigns to promote Thai tourism around the world. These campaigns are Visit Thailand Year (1982 and 1987), Amazing Thailand (1998–1999), and Discover Thainess (2015–2016) [2]. In 2019, Thailand has launched #UnboxThailand campaign to promote "new travel spots—not just in Bangkok but the rest of Thailand" [3]. Therefore, Thailand's hotel industry got full advantages from these arrivals by generating 61% of their revenue from room occupancy. However, how do these hotel's rooms attract and serve the tourists in the aspect of providing the best design and ambience in their respective premises? As we know, one of the current trends of tourism is they highlighted the ecotourism aspect in tourism industry. This is due to "people care about the CO_2 emissions they're causing and how they can travel in greener ways" [4]. So, how they built and enhance the beauty of their hotels? Are they using any materials that

M. F. B. A. Rani (✉)
Department of Architecture, Faculty of Built Environment and Surveying, Universiti Teknologi Malaysia, UTM Skudai, Iskandar Puteri, Johor, Malaysia
e-mail: muhammadfaizal.ar@utm.my

P. S. D. S. N. I. M. K. B. H. Daud
Department of Architecture, Faculty of Built Environment Building, University of Malaya, Kuala Lumpur, Malaysia
e-mail: nimk@um.edu.my

A. Sayigh (ed.), *The Importance of Wood and Timber in Sustainable Buildings*, Innovative Renewable Energy, https://doi.org/10.1007/978-3-030-71700-1_2

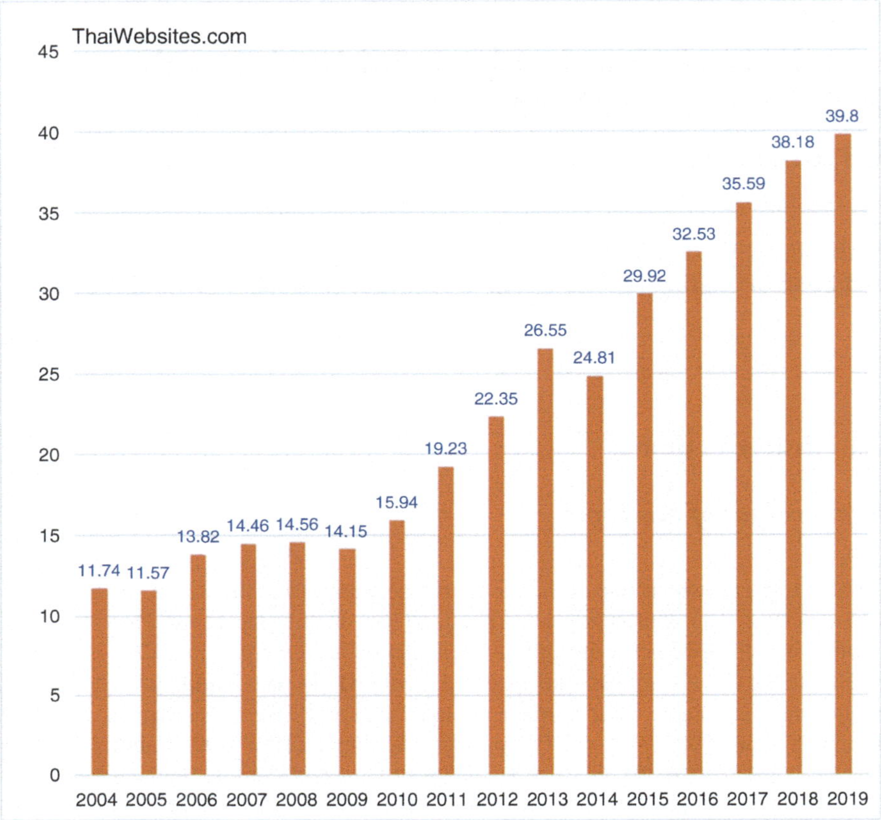

Fig. 2.1 Yearly tourist arrivals between 2004 and 2019 from Vanhaleweyk (2020) [1]

can sustain the environment? There are seven sustainable construction materials in the market, namely, straw bales, bamboo, recycled plastic, wood, rammed earth, ferrock, and timbercrete [5]. In this regard, wood is the most popular material used in Thailand, and it can be easily found in Thai forest. Forest area in Thailand, which is representing about 30.92% of the total land area in Thailand, received annual rainfall ranges from 600 to 3800 mm. Due to this climate, it is very conducive for the tree growth especially in northern region of Thailand (Fig. 2.2). Moreover, the demands for each wood products vary and "some for domestic consumption while others for exportation" ([6]: p. 16). In Thailand, the teak wood is the most popular material for art and construction. This is because the teak wood has many special characteristic such as "lightness with strength, stability, durability, ease of working without cracking and splitting, resistance to termites, resistance to fungi, resistance to weather and non-corrosive properties" [7]. In Thai people culture, they believe that certain species of wood have special importance such as magical properties, protection, healing power, and fertility. In this study, the role of wood in sustainable building will be observed and analyzed based on how these wood have been applied.

Fig. 2.2 The distribution of forest area in Thailand from the Royal Forest Department (2020) ([6]: p. 8)

2.2 Objective

To have an overall view of the application of wood as architectural ornaments in Thailand.

2.3 Scope

The scope of this article only focuses on five-star hotel buildings in Chiang Mai. Five-star hotels were determined to avoid bias during the evaluation. Furthermore, five-star hotels have specific criteria to represent their brand identity.

Chiang Mai province has been selected for these case studies because of its location, which is in the Northern Green Region of Thailand (Fig. 2.3). Chiang Mai is one of the most important provinces in the northern region. It is the center of Lanna

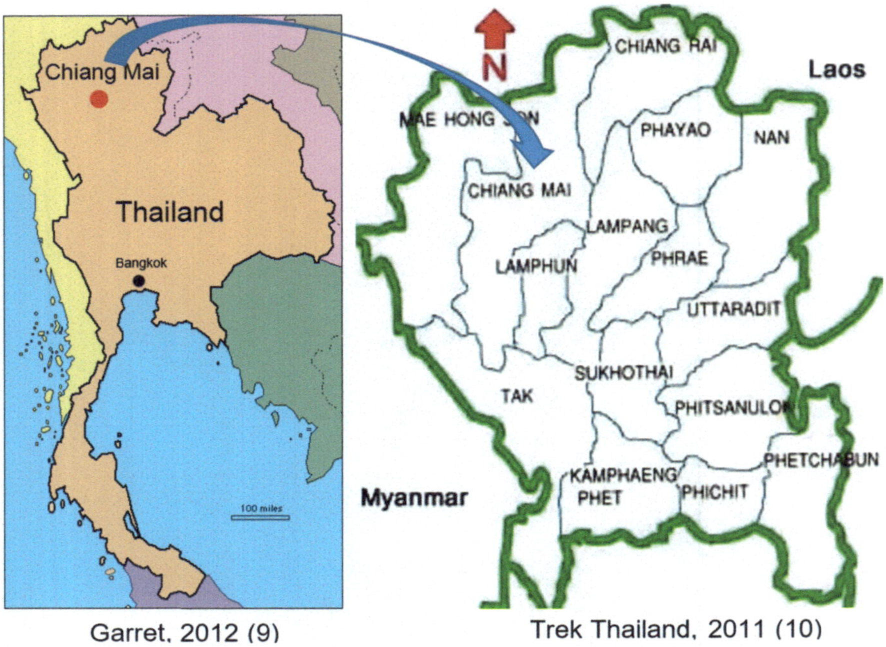

Fig. 2.3 Showing Lanna civilization within the Chiang Mai Province. Garret (2012) [9]. Trek Thailand (2011) [10]

civilization and rich culture that has always been represented through the form of architecture, clothing, and lifestyle [8].

"Chiang Mai is the most award-winning province in Thailand" [11], and also, "the area has many main cultural attractions in Thailand" [12]. Hence, Chiang Mai was named the top city in Asia, as well as being ranked second as one of the world's best cities in the Travel + Leisure World's Best Awards 2016 reader's survey [13].

The hotel lobby area was selected to be observed due to some issues, such as hotel guest privacy, safety, and comfort. Besides, "the lobby plays a crucial role in branding and creating the hotel's desired atmosphere" [14]. The hotel lobby can create the first impression for the guest, and the beauty and attractiveness of the lobby design can influence hotel guests in their decision-making, particularly whether they would want to stay or otherwise [15].

2.4 Methodology

Sampling Design and Data Collection

This article involved direct observation at seven five-star hotels in Chiang Mai province; all these hotels were selected by applying purposive sampling technique. The hotels were selected based on the specific criteria of being a five-star hotel in Chiang Mai only (Table 2.1).

Table 2.1 List of the five-star hotels in Chiang Mai province

No.	Name of the hotel
1	Le Meridien Hotel Chiang Mai
2	DusitD2 Chiang Mai
3	Dhara Dhevi Chiang Mai
4	Four Seasons Resort Chiang Mai
5	Ratilanna Riverside Spa Resort Chiang Mai
6	Shangri-La Hotel Chiang Mai
7	The Anantara Chiang Mai Resort

Measurements

The hotel lobby area was evaluated by using hotel checklist (Table 2.2) during observation process. The hotel checklist was applied to examine the interior architectural space of the selected hotel lobbies "delimited by three planes-a floor, a wall, and a ceiling" [16]. At each plane, any application of the wood was documented by either photographs or sketches. The application of the wood was categorized into building structure and building nonstructure. For building structure, the items include (1.1) roof structure, (1.2) columns, (1.3) beam, (1.4) support structure, (1.5) walls, and (1.6) flooring. For building nonstructure, there are (2.1) doors, (2.2) windows, (2.3) ventilation, (2.4) ceilings, and (2.5) decorative components (lighting, furniture, human objects, animal objects, religious objects, sculptures, wood carving panels, accessories, partition wall). The total number of wood applied in these lobby area was analyzed according to size, form, and function.

Results Discussion

From Table 2.2, only 18 out of 19 items appeared in the lobby with various combinations of application categories and different styles. These show that the emphasized wood applications are more to enhance the sustainability concept of the hotel architecture and lobby interior.

The finding (Fig. 2.4) revealed that there are three aspects of application that could be discussed to show the role of wood as architectural ornaments in current sustainable building in Thailand including the sizes, the forms, and functions:

(a) Size

There are various sizes of the application of wood showed in the hotel lobby area. However, there were more medium- and small-scale wood objects applied. The given reason was that medium- and small-scale objects are always ready and easy to produce and they are widely used and save cost and easy to install. This was confirmed by the finding that refined decorative features in Thai architecture are becoming less sophisticated in modern context, due to lacking of talented workmanship [17].

Table 2.2 Hotel checklist showing the application of wood in two main categories

No.	Hotel's Name	Building Structure						Building Non-Structure												
												Decorative Components								
		Roof Structure	Columns	Beams	Roof support structure	Walls	Flooring	Doors	Windows	Ventilation	Ceilings	Lighting	Furniture	Human Objects	Animal objects	Religious objects	Sculptures	Wood carving panels	Partition wall	Accessories (vase etc.)
1	Dhara Dhevi Chiang Mai	√	√	√			√			√	√	√	√		√	√		√		√
2	Dusit D2 Chiang Mai											√	√							√
3	Four Seasons Resort Chiang Mai	√	√	√			√						√			√				√
4	Le Meridien Hotel Chiang Mai												√							√
5	Ratilanna Riverside Spa Resort Chiang Mai					√		√	√		√	√	√		√	√	√			√
6	Shangri-La Hotel Chiang Mai	√	√	√	√						√		√					√	√	√
7	The Anantara Chiang Mai Resort						√						√							√

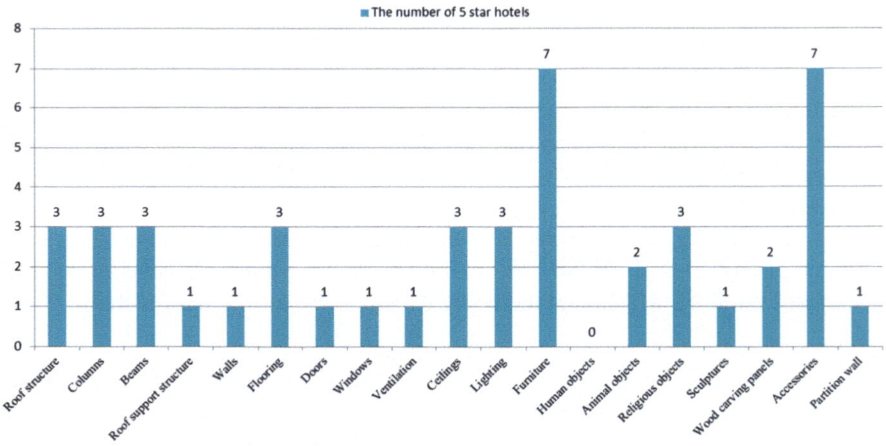

Fig. 2.4 The number of five-star hotels that apply wood in the lobby area in Chiang Mai province

Most of the small-scale objects are made as accessories item (such as small vase, candle stand, food container (Fig. 2.22a), Lanna Buddhist worship offerings (Fig. 2.22b), and traditional musical instrument).

Meanwhile, medium-sized objects include doors (Fig. 2.5), windows (Fig. 2.6), ventilation (Fig. 2.7), ceilings (Fig. 2.8), lighting (Fig. 2.9), furniture (Fig. 2.10), animal objects (Fig. 2.11), religious objects (Fig. 2.12), sculptures (Fig. 2.13), wood carving panels (Fig. 2.14), and partition wall (Fig. 2.15).

Lastly, large-sized objects include the following:

- Roof structure (Fig. 2.16)
- Columns (Fig. 2.17)
- Beams (Fig. 2.18)
- Roof support structure (Fig. 2.19)
- Walls (Fig. 2.20)
- Flooring (Fig. 2.21)

(b) Form

All wood can be found applied in various forms, classified into the following:

- Fauna

– There are three type of animals: elephants (Fig. 2.11d), swans (Figs. 2.11a and 2.19), and bird (Fig. 2.14b). In Thailand, elephants are considered the most favored animals to use for decoration, either for the interior or for the exterior of the buildings. This is because the elephant is a symbol of tradition, history, royal, fortune, and superstition. Furthermore, elephants are closely related with Queen Maya of Sakya, Buddha's mother. The elephant also is a symbol of mental strength; due to this symbol, the elephant is always associated with temple construction and warfare. Meanwhile, the swan is a symbol of divine spirit, perfect union, balance, and life. Usually, swan is applied at the roof ridges of the temple. From this study, the unknown bird motif might be presenting the common bird that can be found in Thailand.

- Flora

– In this study, unknown flower and leaf motif were applied. These motifs were applied at wood carving panels (Fig. 2.14a, b).

- Mythical animal creature

– There are two mythical animal creatures produced from the wood, and these were naga and garuda. This naga (Fig. 2.11b, e) and garuda (Fig. 2.11c) were taken from the Himmaphan forest as detailed in Traiphum. Traiphum, also known as The Three Worlds, represents a Hindu-Buddhist system of hierarchical layers. Inside the Traiphum contains four main continents and seven mountains and seven seas, which are surrounding the mythical Himalayan and Mount Sumeru [18].

(a)

(b)

(c)

Fig. 2.5 Doors at (**a**–**c**) Ratilanna Riverside Spa Resort Chiang Mai. (**a**) Wooden doors behind the reception counter. (**b**) Door handle at the main entrance. (**c**) Door handle

Fig. 2.6 Windows at both side of the lobby, Ratilanna Riverside Spa Resort Chiang Mai

Fig. 2.7 Ventilation at wooden railing at Dhara Dhevi Chiang Mai

(a)

(b)

Fig. 2.8 Ceilings at (**a**) Dhara Dhevi Chiang Mai, (**b**) Shangri-La Hotel Chiang Mai, and (**c**, **d**) Ratilanna Riverside Spa Resort Chiang Mai. (**a**) Buddha image on the lotus position at four main corners, representing four main continents. Single *deva* representing subcontinents. (**b**) Wooden panels were applied on the ceiling above the reception counter. (**c**) Wooden frames were applied around silverwork with *mandala* motif. (**d**) Ceiling was decorated with *ma tang mai*-like structure

(c)

(d)

Fig. 2.8 (continued)

- Buddhist cosmology
  - At Dhara Dhevi Chiang Mai, Buddhist cosmology was applied by having a prasat (Fig. 2.12a) structure at the middle of the lobby area. This prasat represents Mount Sumeru (a mythical mountain, the center of the universe which is surrounded by the four continents and seven mountains and seven oceans). Located on Mount Sumeru is Tavatimsa heaven ruled by the god Indra and surrounded by four kinnara (Fig. 2.12b) representing the four main continents. Kinnara is a male divine musician, half-human and half bird in form, who lives in the Himmaphan forest.
- Lanna Buddhist worship offerings
  - In Lanna Buddhist culture, the making of worship offerings for the temple and religious ceremonies is very important in their life. Making these worship offerings will inspire Lanna people to have faith and follow Thamma teaching. They also believe that making offerings to the Triple Gems will give them merit and prosperity in their life. This merit will take them to the restful life

Fig. 2.9 Lighting at (**a**) Dhara Dhevi Chiang Mai, (**b**) DusitD2 Chiang Mai, and (**c**) Ratilanna Riverside Spa Resort Chiang Mai. (**a**) *Cho fa* can be found in Lao and Thai temples and palaces. It is usually installed at the top end of the roof with decorative ornaments. (**b**) Pendant light with onion shape. (**c**) Stand light with *pahn* shape

Fig. 2.10 Furniture at (**a**) Dhara Dhevi Chiang Mai, (**b**) DusitD2 Chiang Mai, (**c**) Four Seasons Resort Chiang Mai, (**d**) Le Meridien Hotel Chiang Mai, (**e**) Ratilanna Riverside Spa Resort Chiang Mai, (**f**) Shangri-La Hotel Chiang Mai, and (**g–i**) The Anantara Chiang Mai Resort. **a**) Furniture with triangle pillows. (**b**) Wooden furniture was mixed and matched with other types of furniture. (**c**) Reception counter with *buranakata* motif. (**d**) Wooden furniture was almost applied for the whole area. (**e**) Wooden furniture was made in the form of traditional palanquin (**f**) Modern wooden cabinet. (**g**) Lanna traditional kitchen pantry cabinet. (**h**) Lanna traditional low cabinet with Chinese influence. (**i**) Minimalist style of wooden coffee table

(d)

(e)

(f)

Fig. 2.10 (continued)

(g)

(h)

(i)

Fig. 2.10 (continued)

after they die. From this study, only two forms were found, hsun-ok (Fig. 2.22a) and soom dock (Fig. 2.22b).

- Buddhist temple architecture/structure

– Buddhist temple architecture is applied in terms of the roof structure (Fig. 2.16b), roof support structure (Fig. 2.19), ceilings (Fig. 2.8a,c,d), and cho fa (Fig. 2.9a).

- Old palace

Fig. 2.11 Animal objects at (**a**–**c**) Dhara Dhevi Chiang Mai and (**d**, **e**) Ratilanna Riverside Spa Resort Chiang Mai. (**a**) Swan was selected to decorate the space. (**b**) *Naga* attached to the column. (**c**) *Garuda* at the center of the lobby area. (**d**) Elephant form was applied at the garden. (**e**) *Naga* in front of the lobby main entrance door

  - Dhara Dhevi Chiang Mai has imitated Mandalay Palace roof structure (Fig. 2.16c) from Burma.

- Ancient house

  - Ratilanna Riverside Spa Resort Chiang Mai has applied one part of the Lanna ancient-style wall structure. This hotel have applied *Pa-Ka-Na-Ta-Pa* wall which allowed people from inside the house to see out through the holes (Fig. 2.20).

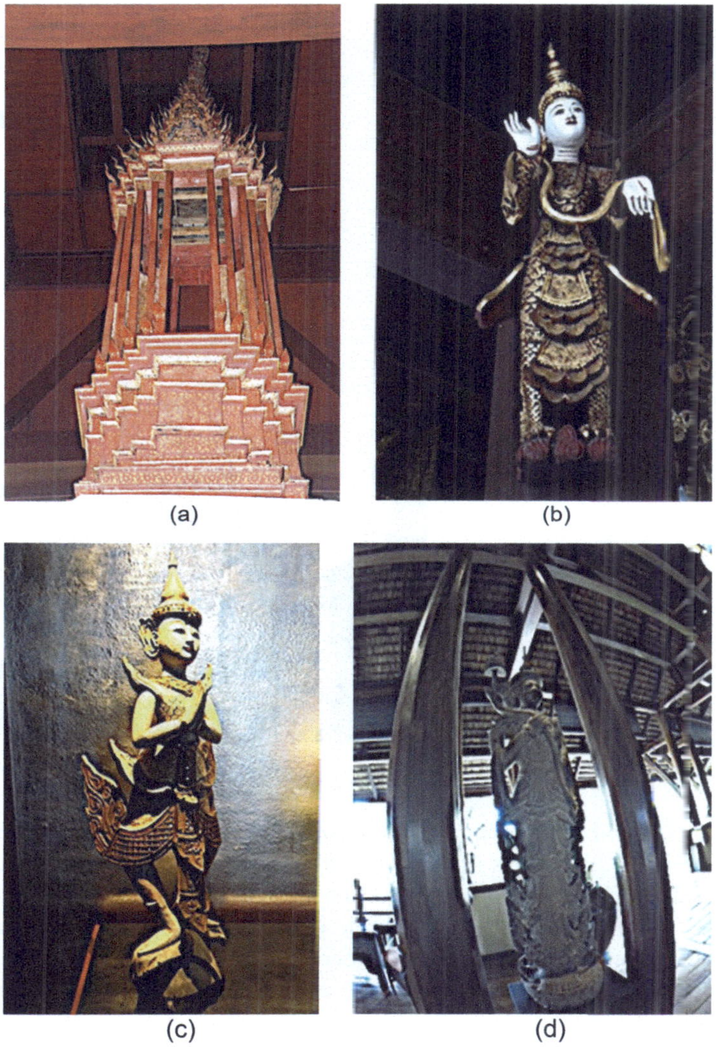

Fig. 2.12 Religious objects at (**a**, **b**) Dhara Dhevi Chiang Mai, (**c**) Ratilanna Riverside Spa Resort Chiang Mai, and (**d**) Four Seasons Resort Chiang Mai. (**a**) *Prasat* at the center of the main lobby area, representing Mount Sumeru. (**b**) *Kinnara* was located at four main corners around *prasat* structure. (**c**) *Deva* in Burmese art style. (**d**) *Deva* in Burmese art style at the entrance leading to the hotel garden

Fig. 2.13 Sculpture at Ratilanna Riverside Spa Resort Chiang Mai

(c) Function

In this study, wood is more applied as decorative items as compared to building structure (Fig. 2.23). This scenario might be because the wood is more easily produced into small- and medium-scale objects as compared to large-scale objects. These small- and medium-scale objects can be produced through low technology, low budget, and short time period and can be made in high numbers. Furthermore, the demand for this small- and medium-scale objects is higher in the market.

2.5 Conclusion

In conclusion, the application of wood is dependent on several factors—the architect or interior designer, demand, and the supply chain. As we know, each architect or interior designer has their own ideas and thoughts. This differences of ideas will determine which architect or interior designer has strong design flair and prefers to use wood in their design. Meanwhile, it is very pertinent for the architect or interior designer to follow the current trend to make sure that their design is as good as other

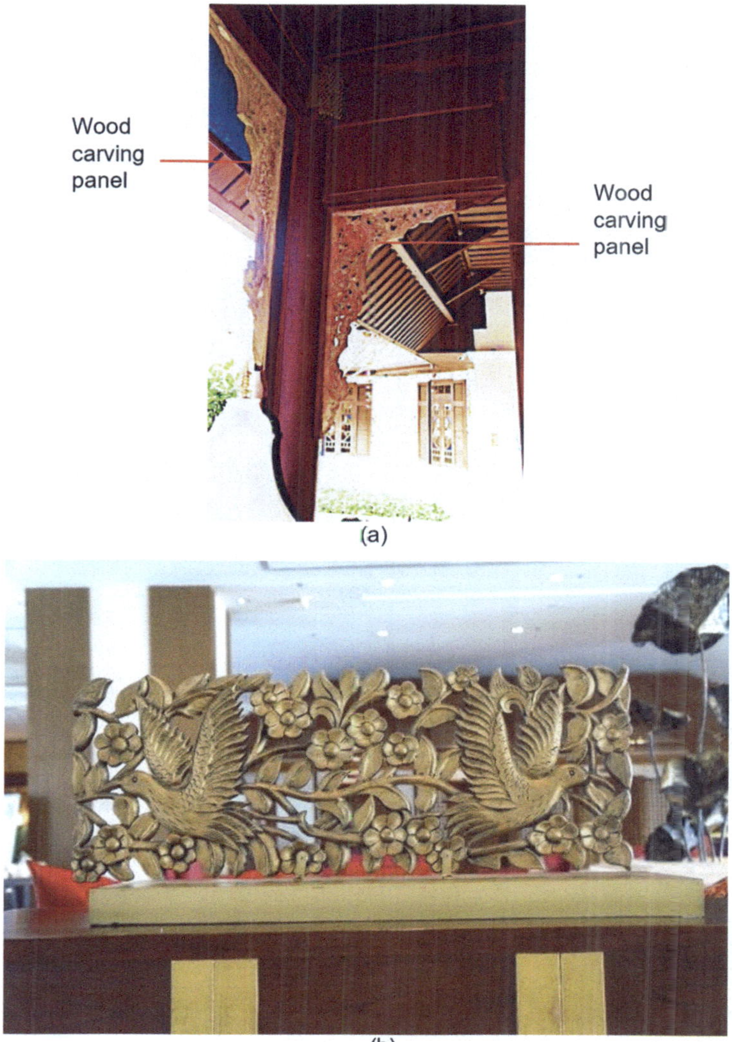

Fig. 2.14 Wood carving panels at (**a**) Dhara Dhevi Chiang Mai and (**b**) Shangri-La Hotel Chiang Mai. (**a**) Wood carving panel with floral motif was attached to the main column as a decorations. (**b**) Wood carving panel with flora and fauna motif

designers. Therefore, once again, the decision to apply wood has become another issue of either it is relevant or not for the selected project.

In terms of demand, there are two types of clients. The first type of client just want to buy the ready-made items such as decoration object (small vase, sculpture, musical instrument, candle stand, wood carving panel, furniture, etc.). The second type of client is willing to have custom-made items for certain purpose. This included to build the building structure such as roof structure, columns, beams, and others. These custom-made items will cost more and limit clients willing to pay for

(a)

(b)

Fig. 2.15 Partition wall at Shangri-La Hotel Chiang Mai. (**a**) Partition wall was installed at the lift lobby area to create the transition area from public area to semipublic area. (**b**) Partition wall was applied to reduce the glare and heat effect

(a)

(b)

(c)

Fig. 2.16 Roof structure at (**a**) Shangri-La Hotel Chiang Mai, (**b**) Four Seasons Resort Chiang Mai, and (**c**) Dhara Dhevi Chiang Mai. (**a**) Roof structure was applied at the lift lobby area. (**b**) Roof structure applied at Lanna Buddhist temple. (**c**) Roof structure imitates Mandalay Palace roof structure from Burma

Fig. 2.17 Columns at (**a**) Four Seasons Resort Chiang Mai, (**b**) Dhara Dhevi Chiang Mai, and (**c**) Shangri-La Hotel Chiang Mai. (**a**) The column shape imitates Thai Buddhist temple column style. (**b**) There are two types of columns in this lobby, round and octagon shape. (**c**) Round shape of columns was applied

Fig. 2.18 Beams at (**a**) Four Seasons Resort Chiang Mai, (**b**) Dhara Dhevi Chiang Mai, and (**c**) Shangri-La Hotel Chiang Mai

Fig. 2.19 Roof support structure at Shangri-La Hotel Chiang Mai

Fig. 2.20 *Pa-Ka-Na-Ta-Pa* walls at Ratilanna Riverside Spa Resort Chiang Mai

Fig. 2.21 Flooring at (**a**) Four Seasons Resort Chiang Mai, (**b**) The Anantara Chiang Mai Resort, and (**c**) Dhara Dhevi Chiang Mai

(a) (b)

Fig. 2.22 (**a**, **b**) Accessories items at Shangri-La Hotel Chiang Mai. (**a**) *Hsun-ok*, offering box from Burmese art influence. (**b**) *Soom dock* was used to put the flowers for ordination hall in the temple

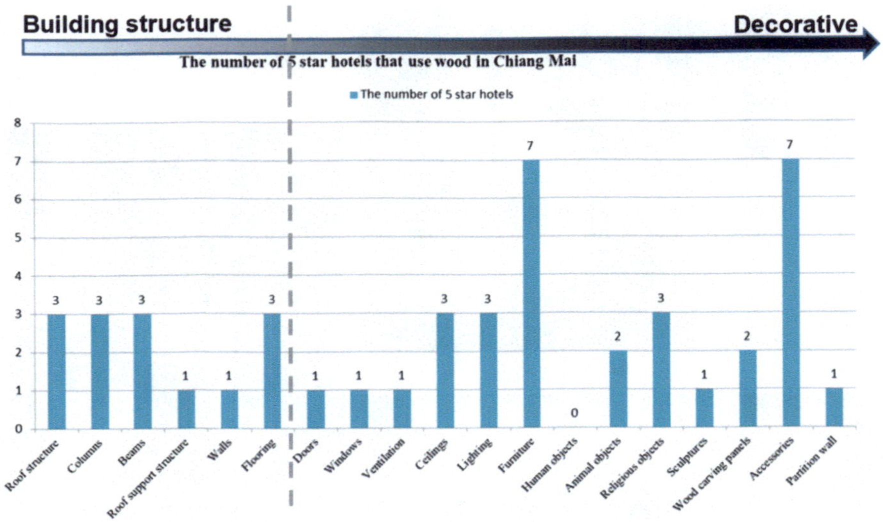

Fig. 2.23 Wood were more applied as a decorations in the lobby area

Fig. 2.24 Type of wood product in the market

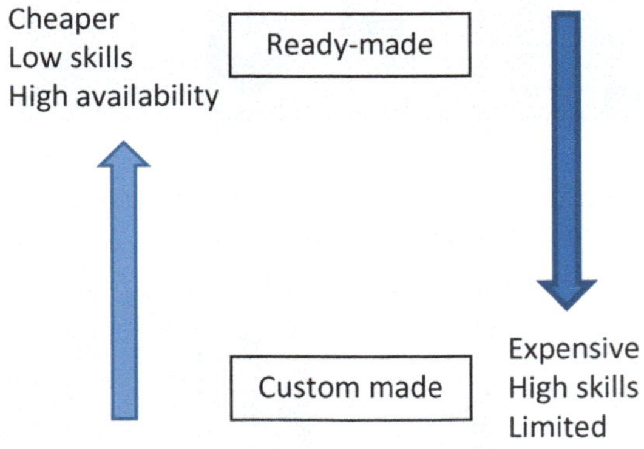

it. Hence, it needs some specific skills in order to produce certain custom-made items (Fig. 2.24).

Based on supply chain operations reference (SCOR) model, wood product can be manageable through five stages: plan, source, make, deliver, and return. In planning, it will involve companies to have major decision of either they want to manufacture a product or component or buy it from the supplier. At this stage, this will involve the capabilities of the involved companies in producing the wood products. This will ensure how the wood product can be sustained in the market. In terms of the source aspect, it is important for the manufacturer to make a great networking with the wood supplier. The manufacturer should know how to organizing the raw materials and components in order to fulfil the clients' needs. In making stage, the manufacturer should be aware with production activities, testing of products, packing, and release. It is very crucial to produce the product based on the timeline because time is money for the clients and manufacturers. The next stage is delivery, which starts from processing clients' query to how to strategize the distribution and transportation options. Therefore, it is important to deliver on what the client needs with precision. There should be no delay in delivering the task. The last stage is return, which is dealing with the product condition. In other words, the manufacturer of the wood product should maintain their quality and ensure their wood product is always top notch. By doing this, the wood can survive and sustain longer in the market.

From this study, Dhara Dhevi Chiang Mai uses the highest number of wood, with 63% each (12 items) into their hotel lobby. This is followed by Ratilanna Riverside Spa Resort Chiang Mai at 53% (ten items), Shangri-La Hotel Chiang Mai at 47% (nine items), Four Seasons Resort Chiang Mai at 37% (seven items), DusitD2 Chiang Mai and The Anantara Chiang Mai Resort at 16% (three items), and Le Meridien Hotel Chiang Mai at 11% (two items) (Table 2.3).

Last but not least, only two hotels (Dhara Dhevi Chiang Mai and Ratilanna Riverside Spa Resort Chiang Mai) applied wood more than 50% in their lobby area. Dhara Dhevi Chiang Mai, which implemented Mandalay Royal Palace style from Myanmar, had fully used wood almost in every aspect of design in order to show the variety of styles that influenced Lanna culture throughout its history. Meanwhile, Ratilanna Riverside Spa Resort Chiang Mai was trying to show the day-to-day life of the Lanna community in the Wat Gate area. Most of the design elements in this hotel represent the handicraft of Lanna community in that area.

However, the role of wood in enhancing the beauty of the space in sustainable building in Chiang Mai was dependent on the overall concept of the hotel. If the hotel concept is based on the local culture and vernacular architecture, obviously, it will use wood as their main medium for construction such as Dhara Dhevi Chiang Mai and Four Seasons Resort Chiang Mai. If the concept is more into modern design style, they only applied wood at certain part of the lobby area such as at the ceiling (Ratilanna Riverside Spa Resort Chiang Mai), wall (Ratilanna Riverside Spa Resort

Table 2.3 The rank of hotel based on the application of wood in the lobby area

Rank	Hotel name	Percentage	Number of items/19
1	Dhara Dhevi Chiang Mai	63%	12
2	Ratilanna Riverside Spa Resort Chiang Mai	53%	10
3	Shangri-La Hotel Chiang Mai	47%	9
4	Four Seasons Resort Chiang Mai	37%	7
5	Dusit D2 Chiang Mai	16%	3
5	The Anantara Chiang Mai Resort	16%	3
6	Le Meridien Hotel Chiang Mai	11%	2

Chiang Mai), floor (The Anantara Chiang Mai Resort), doors, and windows (Ratilanna Riverside Spa Resort Chiang Mai). Furthermore, some of the hotels prefer to cover up certain structure of the building with wooden material. From this study, column structures at Shangri-La Hotel Chiang Mai and The Anantara Chiang Mai Resort were covered with teak wood panel.

References

1. Vanhaleweyk, G. (2020). *Tourism Statistics Thailand 2000-2020* [Internet]. Bangkok: Thaiwebsites; 2020 [cited 24th August 2020]. Retrieved from https://www.thaiwebsites.com/tourism.asp.
2. Sritama, S. (2014). *PM to launch 'Discover Thainess' campaign* [Internet]. Bangkok: The Nation Thailand; 2014 [cited 4th September 2016]. Retrieved from http://www.nationmultimedia.com/news/business/macroeconomics/30250850.
3. Marketing Interactive. (2019). *Tourism Authority of Thailand 'unboxes' various tourist experiences with new campaign* [Internet]. Malaysia: Lighthouse Independent Media; 2019 [cited 24th August 2020]. Retrieved from https://www.marketing-interactive.com/tourism-authority-of-thailand-unboxes-various-tourist-experiences-with-new-campaign.
4. Huber, C. (2019). *Current trends in the travel & tourism industry* [Internet]. Atlanta: Airsage; 2019 [cited 25th August 2020]. Retrieved from https://blog.airsage.com/blog/current-trends-in-the-travel-tourism-industry/.
5. CRL. (2018). *7 sustainable construction materials* [Internet]. London: CRL Management Limited; 2018 [cited 25th August 2020]. Retrieved from https://c-r-l.com/content-hub/article/sustainable-construction-materials/.
6. Royal Forest Department. (2009). *Forestry in Thailand* [Internet]. Bangkok: Ministry of Natural Resources and Environment; 2009 [cited 25th August 2020], pp. 8 and 16. Retrieved from http://forprod.forest.go.th/forprod/ebook/%E0%B8%81%E0%B8%B2%E0%B8%A3%E0%B8%9B%E0%B9%88%E0%B8%B2%E0%B9%84%E0%B8%A1%E0%B9%89%E0%B9%83%E0%B8%99%E0%B8%9B%E0%B8%A3%E0%B8%B0%E0%B9%80%E0%B8%97%E0%B8%A8%E0%B9%84%E0%B8%97%E0%B8%A2/Forest%20in%20thailand%20eng.pdf.
7. Kaosa-ard, A. (1989). Teak (Tectona grandis Linn.f) its natural distribution and related factors. *Natural History Bulletin of the Siam Society, 29*, 55–74.
8. Angkasith, R. (2007). *Translation of Lanna architecture in modern terms* [Internet]. Hawaii: University of Hawaii; 2007 [cited 4th September 2020]. Retrieved from http://hdl.handle.net/10125/45816.
9. Garret, R. D. (2012). *Chiang Mai, Thailand: Visual cues of globalization* [Image on internet]. 2012 [cited 20th August 2020]. Retrieved from https://worldofrobdotcom.wordpress.com/.
10. Trek Thailand. (2011). *Chiang Mai Map* [Image on internet]. 2011 [cited 20th August 2020]. Retrieved from http://www.trekthailand.net/p1/.
11. National News Bureau of Thailand. (n.d.). *Chiang Mai has won numerous awards as best destination for tourism* [Internet]. [cited 8th October 2015]. Retrieved from http://thainews.prd.go.th/website_en/news/news_detail/WNECO5712050010057.
12. Horwath HTL. (2015). *2015 Thailand hotel industry survey of operations* [Internet]. Asia Pacific: Horwath HTL; 2015 [cited 8th October 2015]. Retrieved from http://thaihotels.org/wp-content/uploads/2015/06/HorwathHTLAnnual-Study—-Thailand-2015.pdf.
13. Travel and Leisure. (n.d.). *World's best cities* [Internet]. [cited 10th September 2016]. Retrieved from http://www.travelandleisure.com/worldsbest/cities#intro.

14. Rutkin, K. M. (2005). *User preference of interior design elements in hotel lobby spaces* [Internet]. Florida: University of Florida; 2005 [cited 31st March 2016]. Retrieved from http://etd.fcla.edu/UF/UFE0010323/rutkin_k.pdf.
15. Countryman, C. C., & Jang, S. C. (2006). The effects of atmospheric elements on customer impression: The case of hotel lobbies. *International Journal of Contemporary Hospitality Management., 18*(7), 534–545.
16. Ashihara, Y. (1981). *Exterior design in architecture*. New York: Van Nostrand Reinhold.
17. Horayangkura, V. (2017). In search of fundamentals of Thai architectural identity: A reflection of contemporary transformation. *Athens Journal of Architecture [Internet], 3* 1), 21–40. https://doi.org/10.30958/aja.3-1-2.
18. Sthapitanonda, N., & Mertens, B. (2006). *Architecture of Thailand: A guide to traditional and contemporary forms*. London: Thames & Hudson.

Chapter 3
Wood Handicraft in the Traditional Architecture of Yemen: Current Dangers and Sustainability Issues

Khaled A. Al-Sallal

3.1 Introduction

Wood or timber has always been a fundamental building material in the history of Yemeni architecture. Before the discovery of cement and introduction of modern construction materials (such as concrete and steel) to the Yemeni building market, wood and timber were used widely to build structures that support buildings roofs, intermediate floors, stairways and bridges. They were built as frames led horizontally to carry the static and dynamic loads of the building and transfer them vertically to the thick masonry bearing walls built of stones, bricks or mud. Wood was also used to make or carve many architectural elements that have important functions in the buildings other than the building structure, including the making of very elegant artistic woodwork that decorated the ceilings, gates, doors and windows of mosques and private houses or palaces. Figure 3.1 shows a wooden takhrim (sun and privacy screen) installed on the exterior of a window in Sana'a. Large cities like the Old City of Sana'a allocated a considerable space in its traditional market to the woodwork in the wood market, in what is known as the "Al-Munjara" market in Bab Al-Yaman. Tourists go to Al-Munjara to enjoy seeing the splendid craftsmanship of the traditional carpenters and how they turn wood, by carving and engraving, into beautiful artefacts, many of which are made to fit into architecture such as windows, doors, decorated ceilings or furniture.

This chapter presents a broad review about the traditional uses of wood in the Yemeni architecture. After the introduction, it starts by a broad background divided into several subsections: the first one classifies the wood uses and how it functioned in the traditional buildings. Under each classification, a number of specific applications and wood artefacts were identified, the second is about the wood material in

K. A. Al-Sallal (✉)
Professor/Consulting Architect & Construction Manager, Concreto, Vancouver, BC, Canada

A. Sayigh (ed.), *The Importance of Wood and Timber in Sustainable Buildings*, Innovative Renewable Energy, https://doi.org/10.1007/978-3-030-71700-1_3

Fig. 3.1 A takhrim (mashrabiya) made from wood installed on the exterior of a window in Sana'a

general and the factors that define the structural properties and wood quality, the third is about wood and carpentry in Yemen, the fourth is about Yemen's geographical landforms and climates and the fifth is about social life in Yemen. After the background, the chapter describes the Yemeni vernacular architecture in terms of its characteristics, unique forms, urban fabric of the old cities, the sociocultural and bioclimatic influences and the building construction methods and materials. This is followed by a detailed analysis on the traditional wood uses in the structural systems, doors and windows and the decorated ceilings. At the end, the chapter presents a discussion of the present threats and future challenges and proposes a general (preliminary) sustainability framework and lastly the epilogue.

3.2 Background

Classifications of the Traditional Wood Uses

The categories of wood uses/functions in the traditional buildings are shown in Fig. 3.2; they can be described as follows:

1. *As a structure system*: wood was used to create frames that support the structures of the building roofs, intermediate floors, staircases, sanitary duct system and bridges between buildings.
2. *As a reinforcing material*: wood branches, twigs and straw were used to support the roof's coverings of buildings and reinforce the walls of mud/earth buildings and cities' defensive walls.

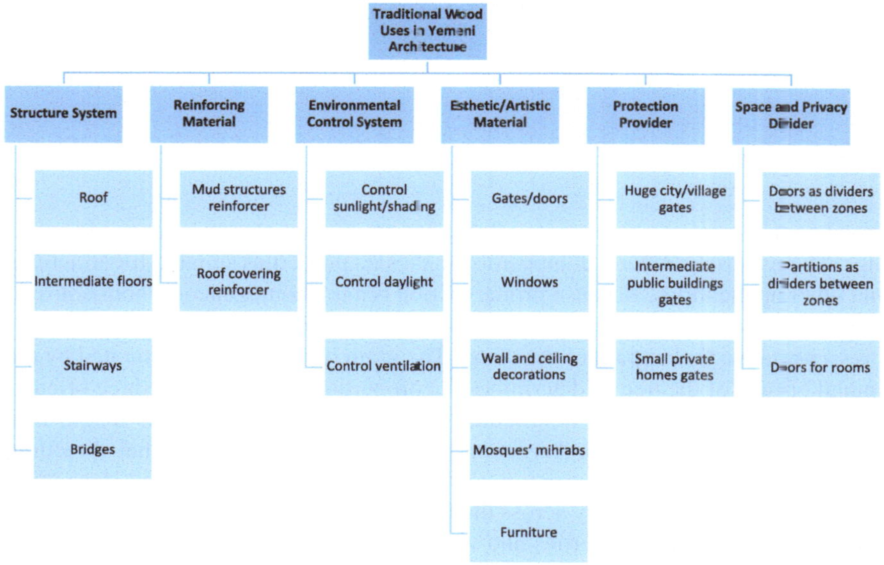

Fig. 3.2 Traditional wood uses in the Yemeni architecture

3. *As an environmental control system*: wood was used to create different kinds of windows and window's components that have different environmental functions: provide view, control sunlight/shading level, control daylighting level and distribution and promote ventilation. The desired function of the window and the influencing changes of the climate in a region (hot or cold, rainy or dry, dusty or clear) were the main determinants of the shape, size, wood kind, method of fixation and operability mechanism.
4. *As an aesthetic/artistic material*: wood handicraft in Yemen is regarded as an elegant art that expresses the Yemeni heritage and strengthens people's ties to their ancient history. As a material, it has high workability (i.e. ease of cutting, nailing) and potential to be shaped into works of art and charming artefacts. The engraved wood artefacts can be found in the traditional gates, doors, windows, wall and ceiling decorations, mosques' minbars (podia) and mihrabs (semicircular niche in the wall of a mosque that indicates the qibla) and furniture. All these traditional elements were made with astonishing craftsmanship and elegance.
5. *As a protection provider*: wood was used to create gates that provide high protection and security for the entries of large cities, small towns, villages, private houses/places and public buildings including governmental buildings, mosques, and shops.
6. *As a space divider and privacy provider*: wood was used to create doors or partitions to provide privacy for zones within the same floor of a building or house. Wood with a fine quality was also used to create screens (i.e. takhrim or mashrabiya) that were installed on the window's external side, which can be found mostly in the large traditional houses or palaces of wealthy people, to

provide a see-through privacy barrier between the house indoors and the public street or neighbouring houses.

Wood in General

Wood demonstrates a unique material that has many characteristics for use in buildings. The material's natural composition makes its properties suitable for structures and other applications. Because primary wood is a nonhomogeneous, non-isotropic material, it exhibits different structural properties depending on the orientation of stresses relative to the grain of the wood. The grain of the wood is generated by the annual growth of the tree. It determines the properties of wood along three orientations: tangential, radial and longitudinal. The grain is parallel to the length of a lumber member because lumber is cut from logs in the longitudinal direction. Depending on where the lumber is cut relative to the centre of a log (i.e. tangential versus radial), properties vary across the width and thickness of an individual member [1]. The structural properties and wood quality depend on several factors that can be outlined as follows (Fig. 3.3):

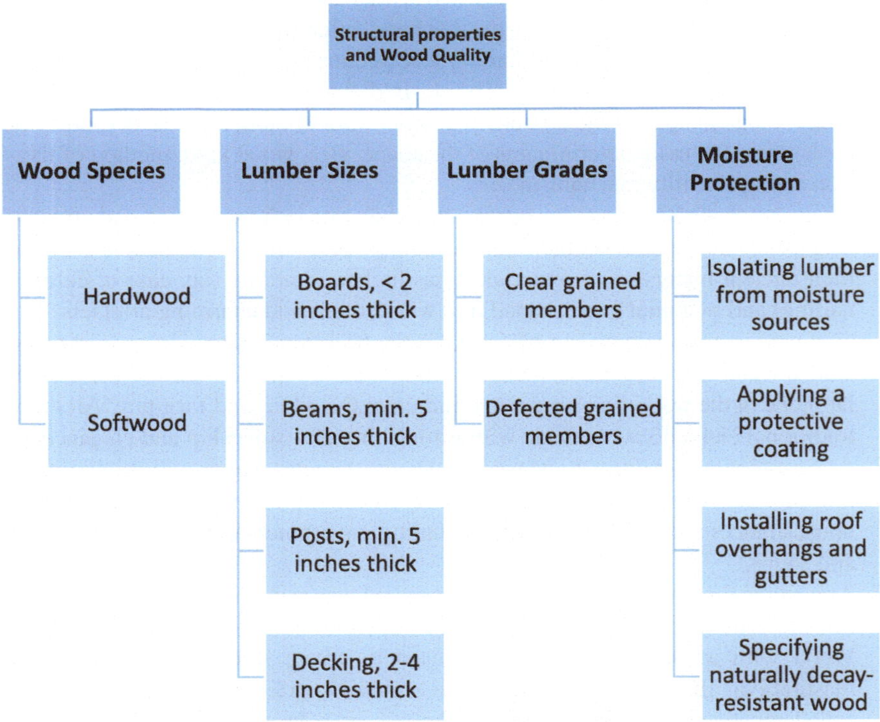

Fig. 3.3 Factors identifying wood structural properties and quality

1. *Wood species*: The various species used in a given locality are a function of the economy, regional availability and required strength properties. Generally, a wood species is classified as either hardwood or softwood. Hardwoods are broad-leafed deciduous trees, while softwoods have needle-like leaves and are generally evergreen. Most structural lumber is manufactured from softwoods because of the trees' faster growth rate, availability and workability.
2. *Lumber sizes*: Wood members are referred to by nominal sizes depending on its application or use. Boards are less than 2 in. thick. Beams are a minimum of 5 in. thick, with the width at least 2 in. greater than the thickness dimension. Posts and timbers are a minimum of 5 in. thick. Decking is 2–4 in. thick and loaded in the weak axis of bending for a roof, floor or wall surface.
3. *Lumber grades*: Lumber is graded based on standards that consider the effect of natural growth characteristics and defects, such as knots and angle of grain, on the member's structural properties. Defects can affect the overall strength of the member relative to a perfect, clear-grained member without any natural defects.
4. *Moisture*: Moisture is a primary factor affecting the durability of lumber. When wood is subject to moisture levels above 20%, decay begins to set in. Fungi, which feed on wood cells, require moisture, air and favourable temperatures to survive. Therefore, it is important to protect wood materials from moisture. This can be done by isolating lumber from moisture sources (such as ground contact), applying a protective coating (e.g. paint, water repellent), installing roof overhangs and gutters or specifying naturally decay-resistant wood.

Traditional Woodworking

The traditional wood handicraft in Yemen has been one of the oldest and finest in the Arab region. Carpentry and engraving on wood are an art and science that require high skills passed from one generation to another in Yemen, especially in the carving and engraving old market of wood in what is known as the "Al-Munjara" market in Bab Al-Yaman quarter of Sana'a. Amongst the most important places in which this splendid heritage thrived were in the old Al-Munjara markets of Wadi Hadramout cities and towns (such as Shibam and Dar Al Hajar), Sana'a, Dhamar, Al-Hudaydah and Zabid. In the Old City of Sana'a, many factors played a role that promoted the historically famous carpentry craft, amongst which is the presence of highly skilled artists and craftsmen who produced pleasing and harmonious artefacts that decorated the interiors and exteriors of the buildings and homes. Their splendid craftsmanship turn deaf wood into works of art and charming beauty [2].

The Tanab is one of the finest and most preferred wood in Yemen because it can last for hundreds of years without being damaged by termite or affected by climatic changes. The Tanab wood comes from the local Yemeni Tanab tree; which is named scientifically as "Cordia Africana." Directorate Maghrab Ans in Dhamar Governate is famous for its premium quality of the Tanab. Besides its strength and durability, the Tanab has high workability and thus it is an excellent wood for carving and cutting, for its ability to resist twist and possibility of damages that could be caused by the shaping machinery. Al-Talh is another wood that was used in the manufacture of strong gates. It is not desirable as the Tanab though, due to its ability to fracture during the shaping process. Therefore, carpenters do not carry out work on Al-Talh wood unless it is green. Al Talh is also considered as the best wood to manufacture ploughs. As for the redwood and plum species, they were used when wood lathing was needed in the making of traditional furniture that is decorated and embroidered with silver.

In wadi Hadramout, the Sedr (Ziziphus spina-chris) and Ethel (tamarix) can be seen in the woodworking of the traditional buildings. The Sidr wood was used frequently in the traditional carpentry because of its high quality and distinctive properties [3]. It is a heavyweight wood with high qualities of hardness and longevity and has a high resistance against fungi and intrusion of insects. Nowadays, it is very expensive wood. The Ethel is a hard wood and was used for carpentry and making engraved artefacts.

The traditional woodworking process comprised several stages from obtaining the raw material till making the final artefact [4, 5]. It can be outlined as follows:

1. Obtaining the raw wood from the plantation suppliers
2. Drying the wood in several stages
3. Selecting the wood suitable to the needed artefact
4. Cutting the wood according to the needed pieces
5. Shaping the pieces to become standard size forms or members
6. Designing and forming the friezes' templates
7. Retracing the templates' lines on the wood
8. Carrying out the carving and engraving process
9. Sanding and polishing
10. Transporting and delivering product orders.

The dangers that are encountered in sustaining this beautiful Yemeni handicraft are enormous today. The future can be worse if it is not safeguarded. At the moment, the most devastating danger comes from the ongoing war that resulted in a collapsed economy. One can find in the Internet some resources that describe the shattering effect of the war on the economy and how the traditional wood handicraft in Yemen was affected badly [4, 5]. Section 3.5 of this chapter explains more details.

Woodworking at the Present Time

The kinds of wood that are mostly available today in Yemen include meranti, beech (zan) and teak. Their general properties and uses can be explained as follows:

- *Meranti wood*: Meranti is a versatile hardwood with many subspecies, sources and purposes. Philippine or lauan mahogany is another somewhat generic term for meranti that is harvested and is available in abundance on a worldwide basis [6]. Meranti comes in two forms: light red meranti or dark red meranti. With its deep brownish-red tone, highly interlocked grain and high degree of resistance to warping or twisting, meranti wood is perfect for moulding, structure, furniture, cabinets and veneers [7].

- *Beech wood (Al Zan wood)*: It is widely used for furniture framing and carcass construction, flooring and engineering purposes, in plywood and in household items like plates, but rarely as a decorative wood [8].
- *Teak wood*: Teak (*Tectona grandis*) is a tropical hardwood of large, deciduous tree species in the family Lamiaceae. Teak wood has a leather-like smell when it is freshly milled and is particularly valued for its durability and water resistance. The wood is used for boat building, exterior construction, veneer, furniture, carving, turnings and other small wood projects [9, 10].

Yemen's Geographical Landforms and Climates

Arabia Felix was the Latin name previously used by geographers to describe South Arabia or what is now Yemen [11]. It literally means Fertile, Happy or Lucky Arabia [12]. It was also described in Ancient Greek as Eudaemon Arabia [13]. Felix has the simultaneous meaning of "fecund, fertile" and "happy, fortunate, blessed." Arabia Felix was one of three regions into which the Romans divided the Arabian Peninsula: Arabia Deserta, Arabia Felix and Arabia Petraea. The Greeks and the Romans called Yemen "Arabia Felix" [14, 15].

Amongst the important aspects that made Yemen such a happy or blessed place are its diversified geographical landforms and climates. Yemen's geographical landforms consists of the following:

- High mountain chains that are beautifully contoured by villagers to create terraced gardens for growing crops, for example, the finest and most famous coffee "Arabica Coffee." Historically, the Yemeni grape was mentioned as the best, and there is a famous Arab aphorism "we never got to the Sham's dates or Yemen's grape" to express disappointment in not obtaining the thing that we wish to have signifying its precious value. The western and central highlands include the highest mountain in the Arabian Peninsula Jabal An-Nabi Shu'ayb with elevation of 3666 m (or 12,028 ft). The climate of the highlands is pleasant and comfortable all day long in most days of the year with some cold nights in the winter.
- Huge fertile valleys like the famous Wadi Dawan in Hadramout that produces one of the finest honeys in the world. Shibam City has a hot desert climate with an average temperature of 28 °C.
- A very long coastline that extends along the Arabian Sea, the Gulf of Aden and the Red Sea over a distance of 1906 km (or 1184 mi) with a rich marine resources including one of the best fishery places for tuna and other seafood species. The climate in the coastlines is hot and humid in the summer, while it is comfortable and pleasant in the winter.

- Many islands full of natural resources including the legendarily island of Socotra that has been described as a paradise or the most alien-looking place on Earth [16, 17]. Socotra Island was recognized in 2008 as a UNESCO World Heritage Site [18]. The climate of Socotra is classified as a tropical, desert climate bordering on a semi-desert climate with a mean annual temperature of over 25 °C [19].
- Unique deserts on the east and southeast sides with many remarkable historical sites such as the famous Sabaean temple of Awwām or the Maḥram Bilqis (Sanctuary of the Queen of Sheba) and the Ancient Dam of Ma'rib, both in the Ma'rib Governorate of Yemen.

Social Life in Yemen

The Yemenis in general are very socializing people and compassionate in keeping intimate relations with their family members, neighbours, and friends. The very active social life of the Yemeni people necessitates welcoming many guests into their homes. They strengthen their social relations by celebrating special events (e.g., religious feasts, weddings, condolences' visits, etc.) with others in homes, and exchange simple daily or weekly visits with family members and friends. They also have the habit to visit their relatives in their homes and check their needs often after Friday prayers, especially the seniors, women and needy ones such as the aunts, sisters, uncles, and cousins. The need to create a building that accommodates a large number of family members and satisfies their needs for celebrating social events and welcoming many visitors have a strong effect on the form of the house and the way it was constructed. This includes the design of the structure; the house entry, the circulation, and the staircase; the vertical and horizontal distribution of rooms; and the sanitary system and bathrooms. The unique tower house provided a very suitable form for the people's social life and cultural activities due to its relatively large size and potential for offering multiple levels of privacy zones (the highest are the most private for the family members while the lowest are the least private). Bridges were also created to cross between the top floors of the tower houses of the close relatives (like brothers and sisters).

3.3 Vernacular Architecture

Yemen has very distinctive regional architectures (Fig. 3.4). The most unique features are the slender vertical forms of the tower houses and the attractive designs of its façades and windows. It can be easily recognized by its strong character and deep cultural meaning while blending beautifully with the urban fabric and the surrounding nature, such as mountains or desert.

Fig. 3.4 Vernacular architecture: the high-mountain vernacular architecture in Bora village and its harmony with the mountains' contours (left); Lakamat al-Gadi fortress in Jabal Haraz (right); the high-rise architecture of Shibam, Wadi Hadramout (below left); and Al Hajarayn town in Wadi Dawan (below right) [20, 21]

Urban Fabric of the Old City

The author discussed in depth the morphology of the Old City of Sana'a in Yemen, in terms of its street's orientation and the achieved balance between building masses and urban or green spaces and how this balanced morphology helped to offer sites with great opportunities for harnessing the solar radiation needed for winter heating while improving shading and passive cooling in the summer [22–29].

In Yemen's old cities like Sana'a, one can observe a clever urban system composed of housing clusters with access to the public roads and squares from one side while having another access to plentiful gardens from the other side. The gardens of the Old City of Sana'a occupy 20% of the city's area and 42% of each residential quarter; thus, almost every house has a view through its windows into extensive gardens (Fig. 3.5). Such organization provided high integration of the tower houses with the rest of the city's urban fabric, placement of tower houses next to the gardens as the main source of food (grown harvests of vegetables/fruits and small farm animals like chicken or goats). This required to create a sophisticated system of water management infrastructure that runs from underground water wells to cisterns and branching channels underneath the buildings or in between them, which supply water to every housing cluster and garden. The city also had a historical storm water canal "al-Sailah" that protected it against water flooding while keeping the

Fig. 3.5 Aerial view showing the traditional gardens in the Old City of Sana'a (left) and a close-up view of an example of the traditional garden within the residential quarter (right) (26-SDNS, 2011) [30, 31]

infiltration of water into the deep soil, which always maintained sufficient water table for its inhabitants since it was founded 3000 years ago.

Compliance with Conventional Standards

To keep everything functioning properly, each city or village followed conventional standards that were widely respected by its inhabitants and practiced by the builders. These standards were the result of history-long experiences that were passed from one generation to another. They were developed to respond to many factors; some were rooted from the Arab and Islamic cultural background, while others were specific to each location. The general aim was to satisfy the sociocultural needs of the people and respond to the requirements of climate and environment. The people's adherence to these standards resulted in producing a consistent architectural language that gave a unique identity to the place and preserved it for long time.

Environmental Factors

People learned how to respect and exploit the environment for their advantage. Climatic conditions were one of the dynamics that formed its architecture. They created buildings that maintained a certain desired level of tolerability (or protection) of a natural influence based on allowing or preventing its magnitude, direction/distribution or effect. There are countless number of environmental strategies. To mention a few examples:

- Walls constructed by sun-dried mud brick in many locations in Yemen, such as the ones in Shibam, Hadramout, have a higher heat capacity and lower conductivity than concrete, which means they slow the rate of heat gain/loss to/from the buildings' interiors. They are also eco-friendly and cheap because their labour costs are the only real costs involved. Not only does the production of sun-dried bricks involve no polluting emissions, the bricks are also reusable [32]

- Sun shading versus sun exposure, "which one to provide," was dependent on the site, season and wall orientation. This is evident in the design of the window's size, shape, orientation and the addition of sun shaders such as the overhangs or shutters and their operability.
- Methods that exploited the wind accessibility into buildings including windows' distribution, sizes and operability of their shutters to promote cross-ventilation or stack effect.
- Methods that controlled the storm water from entering houses and streaming it into basins for later use and the traditionally developed water-resistant plaster used as topping over the roofing system.

Building Materials

Another dynamic was the building materials (mud, stone, wood, gypsum etc.) that vary from one region to another. This depended on the natural riches of each region and the methods that were developed to obtain them: quarry them from rocks like stone, extract them from earth like mud or cut them from trees like wood. The richness and diversity of Yemen's geography provided a limitless resource to discover and adapt the right building material to the desired certain function (strength, reasonable weight, low heat conductivity, waterproofing, aesthetics' qualities like colour or texture, etc.).

The Traditional Tower House

The traditional tower house is the most famous form that characterizes Yemeni architecture. It can be found in the Old City of Sana'a (the capital of Yemen) and also in many other cities/towns. The tower house of Sana'a is a good archetype of the highland architectural form (Fig. 3.6), a region that can be characterized with moderate summers (day and night) and moderate winter during the day time hours while being cold during the night-time. It was built with many floors (usually more than five; the largest commonly had 7–9 floors) that were often occupied by one extended family consisting of 2–3 generations. The tower house walls are built of 50 cm^2, black, volcanic ashlar stone on the lower levels (i.e. up to approximately 6–10 m above street level) and 40 cm baked exposed brickwork above that. The roof consists of a frame of wooden beams covered by branches topped with earth. The lower floor is frequently double-storey in height and used for keeping animals and for storage. The higher floors are residential quarters and include rooms that are functionally polyvalent and non-specific and rooms that can be used interchangeably for eating, sleeping, recreation and domestic tasks. This flexible use of living space is reflected by the absence of cumbersome furniture.

Fig. 3.6 Two ancient examples of the vernacular tower house of Sana'a

Tower houses with different archetypical forms can also be found in the hotter and more desert regions of Yemen such as the tower house of Shibam, Hadramout. The houses of Shibam are all made out of mud brick, and about 500 of them are tower blocks, which rise 5–11 storeys high [33].

Campaign for the Preservation of the Old Cities

UNESCO and the Yemeni government established the "Campaign for the Preservation of the Old City of Sana'a" in 1984 with the goal to preserve the architectural heritage of the old city. The mission of the campaign was to restore and upgrade the city under the direction of the General Organization for the Preservation of Historic Cities in Yemen (GOPHCY). The campaign was outlined in a UNESCO publication which presented a strategy for conservation. The campaign considered the need not only to preserve and restore buildings but also to revitalize social, commercial, educational and economic aspects. Finance was provided by the Yemeni government with the continued help of foreign aid. In 1988, Sana'a was approved by the World Heritage Committee for inclusion on UNESCO's World Heritage List [26]. The campaign was extended to include other ancient cities of Zabid and Shibam, Hadramout, which were also inscribed like Sana'a on the World Heritage List.

3.4 Analysis of Traditional Wood Systems

Roof Structure

The roof in the Sana'a house depends on a frame structure system that consists of thick tree logs or wooden purlins (around 20 cm in diameter), set 50–60 cm apart with each purlin crossing the room's span between two facing bearing walls and anchored to them (Figs. 3.7 and 3.8). The purlins are covered by branches or twigs perpendicular in direction, on top of which lie layers of finely sifted earth mixed with twigs and straw, wet and compact, up to a thickness of 30 cm.

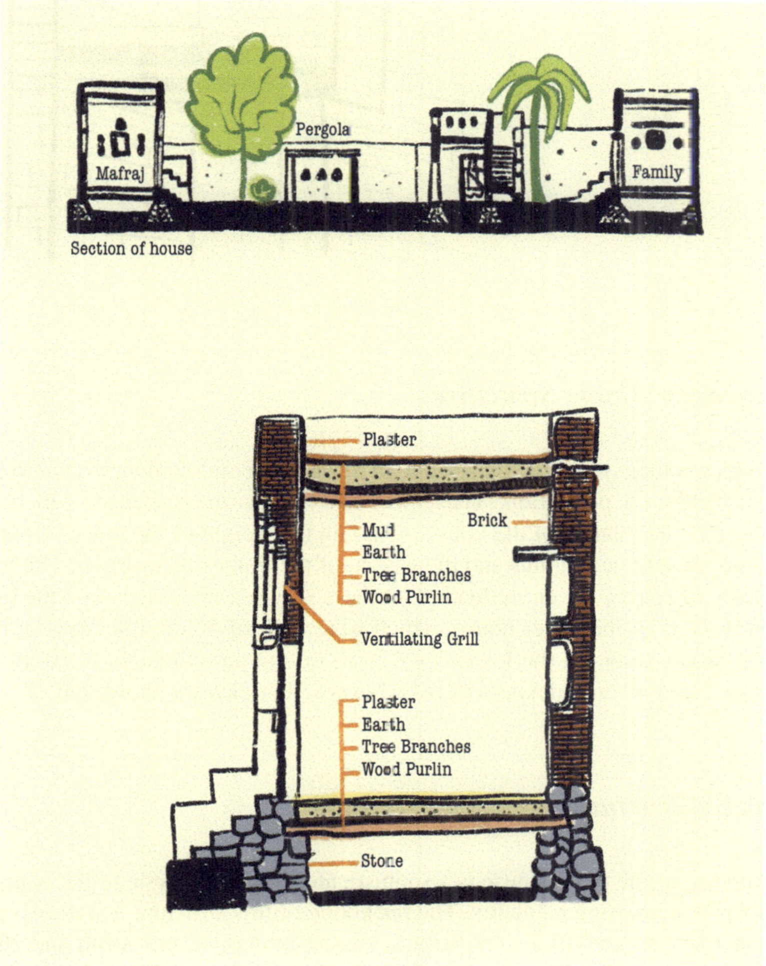

Fig. 3.7 Section showing the structure of the roof and ground floor

Fig. 3.8 3D sketch showing the structure of the roof and intermediate floor

Intermediate Floors Structure

A similar structure system was also applied to carry the middle floors yet with lesser thickness and rain protection. These wooden frames are responsible in bearing everything in the interior of the house including the weight of the house occupants, their furniture and belongings and the weight of the frames themselves. The frames were made to convey the entire load of all these gravitational forces into the rugged structural walls of the tower house, which were built of 50 cm thickness of black, volcanic ashlar stone on the lower levels (i.e. up to approximately 6–10 m above street level) and 40 cm thickness of baked exposed brickwork above that.

Staircase Structure

The staircase of the tower house is a sophisticated structural system that comprises a number of supporting elements. The backbone of this structure is a massive core wall constructed from solid stone with a cross section of 90 cm width and 200 cm length. The core wall is located in the centre of the staircase space that is formed by four boundary walls, as shown in Fig. 3.9. The steps (treads and risers) of the

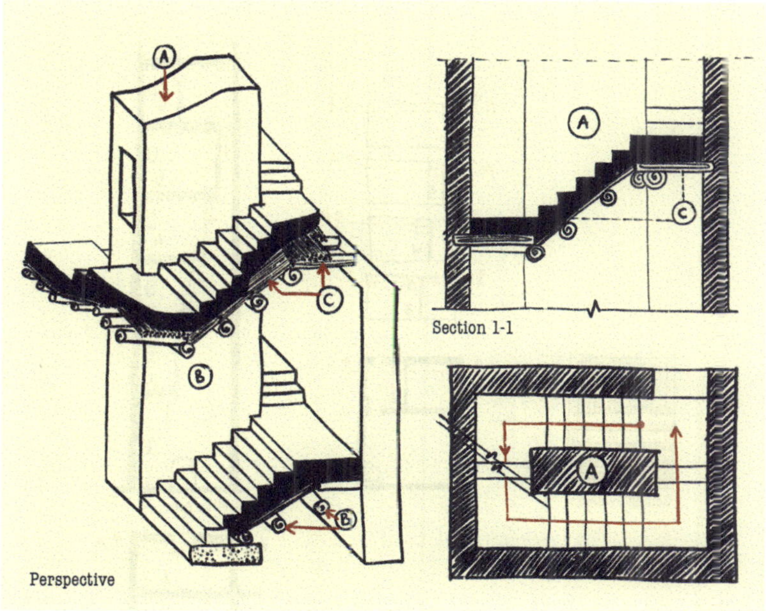

Fig. 3.9 3D sketch, section and plan showing the staircase structure of the traditional tower house of Sana'a

staircase, which are formed by solid stones or sometimes earth, exist around the four sides of the core wall. The loads that are carried by the steps are transferred vertically into twigs and then laterally through thick tree logs or wood purlins (60 cm apart) and then vertically through the core and boundary walls of the staircase. The core stone wall functions not only as a structural member but also as a beautiful decorative element; thus, it is constructed with the good and strong types of stones and are shaped artistically to offer functional slots, shelves and/or wooden cabinets with beautiful carvings; they are used as places for storing or displaying fancy traditional artefacts, glassware, chinaware/ceramic, or even light fixtures.

Traditional Bathrooms and Duct Space Structure

The tower house includes many old-fashioned bathrooms, often one per floor, to serve the large number of family members and also their visitors. The bathroom can be located on the same floor where the rooms exist, or in some cases, they are located in between two floors accessible from an intermediate landing of a staircase. In the past, when these buildings were constructed, modern sewage systems that include the house plumbing, the sewer line from house to septic tank and outlet to sewer pipe did not exist. Sana'a as a city dates back to around 3000 years, and some

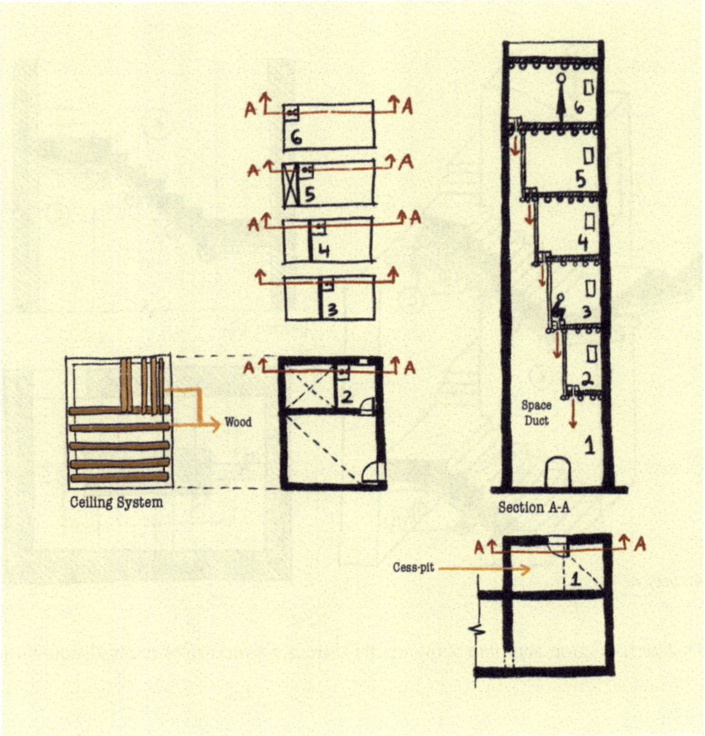

Fig. 3.10 Section and plans showing the structure of the bathrooms and duct space of the traditional tower house of Sana'a

of its heritage houses have existed for more than 600 years, with people still living in them. These houses used an old-fashioned bathrooms and sanitary system. This old system depended mainly on creating a common vertical duct space that extends from the ground to the top floor, with its size diminishing (in one or two directions) as it rises from one floor to another (Fig. 3.10). In a vertical house section, this looks like half a stepped pyramid-like space formed between overlapping floors from one side and the vertical line of the external wall of the house from the opposite side. The construction of this system is made of stonework walls as vertical supports and wooden structure using tree logs or wooden purlins with twigs and earth to create the floor slab. Structurally, it is integrated with the entire body of the house. The structural technique to create the openings on the floors for the duct space to stream between the floors is made by laying down primary row of large logs in one direction (on the large span) crossed by a secondary row of smaller logs that are perpendicular to the primary ones (on the short span).

Fig. 3.11 Examples showing how the tower houses are connected by bridges in Sana'a and Thula

Bridges Connecting Tower Houses

The top floors of some of the tower houses belonging to the same family (like brothers and sisters) are connected by bridges to facilitate movement from one house to another. Especially for women (Fig. 3.11), this is a way to strengthen ties between family members, secure more privacy for women (instead of having them go down to the street) and make the house occupants more comfortable by minimizing their need to climb the very tall, tiring stairs of these houses. The bridge crossing between two houses depended on a frame structure system similar to the one made for roofs (see above). Strong wood purlins laid close to each other were used to ensure high stability for the constructed bridge. Sometimes also, the wooden structure is carried by two corbelled arches constructed by bricks or stones, one from each side of the bridge that connects between two tower houses.

Reinforcing Material

Wood was used in the form of tree branches that are laid over the structural wooden purlins of the roof and covered with a mix of earth and tree twigs (around 20 cm thick), topped with a locally made water-resistant plaster to resist the storm water. The total thickness of this type of traditional roof assembly, including the structural purlins and branches, is around 40 cm. The walls are constructed as thick mud layers (zabour technique) or as rows of mud bricks, stacking on each other. The mud bricks technique is either sun-dried or kiln-dried; each location has a different technique. Near the town of Saada in Yemen, builders use the zabour technique of rammed-earth construction [34, 35] (Fig. 3.12). These walls were reinforced by chopped fine twigs or straw mixed with small gravels. The mud of the mud brick is mixed with chopped straw and water and then spread into simple wooden moulds on the ground to bake hard. The type of earth, which is just the right combination of clay and silt and sand, sets very hard and holds very well with the reinforcing straw, and although a thin outer layer may get washed off during the rainy season, it is basically waterproof [36].

Fig. 3.12 Near the town of Saada in Yemen, builders use the zabour technique of rammed-earth construction. Pascal and Maria Maréchaux

Fig. 3.13 The mud walls of the old cities of Sana'a and Shibam

Much thicker walls using the same mix of materials were used to build the cities' defensive walls (sur) such as the ones surrounding the old cities of Sana'a and Shibam, Hadramout (Fig. 3.13). Large parts of the old Sana'a walls were destroyed especially during the revolution of 1962. These walls were believed by many to be a symbol of the absolute authority of the old regime of Imams and of the total isolation enforced by them; breaking the walls had an important meaning. It represented breaking the prison (or the handcuffs) enforced by the Imams who locked Yemen for centuries. Nevertheless, what caused considerable damage to these walls was the royalists when they attempted to take over Sana'a again in 1968 by attacking it with heavy artillery during the 70 Days War with the Republicans. The damage was said to be not perhaps as much as might have been expected considering the capabilities of modern weapons [26, 37].

Windows

The windows in the Yemeni tower house are usually small in the lower floors, middle size in the middle floors and relatively large in the higher floors (Fig. 3.14). The largest windows in the house can be found at the top floor level, where a small

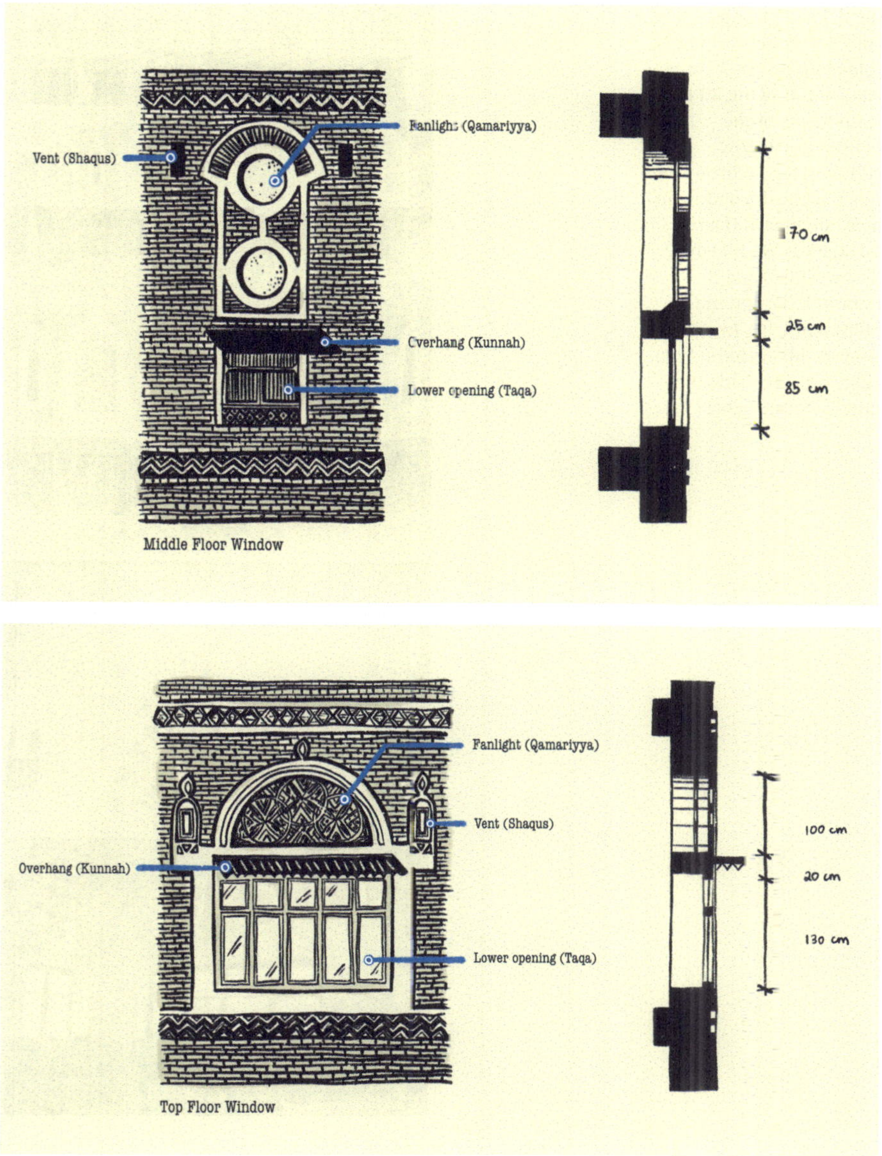

Fig. 3.14 The windows and their components in the traditional tower house of Sana'a: small in the lower floors (above); larger windows in the higher floors (below)

square room, named Tairamana in the Yemeni dialect, is used as an observatory space. The name "Tairamana" comes from Tair in Arabic, meaning bird, to signify it being high-flying like birds. The window has four major parts. Each part has a different function. A window can consist of some or all of these parts, depending on its main function and its location on the façade as well as the plan of the house. Al-Sallal and Cook documented the indigenous window of Sana'a and analyzed its

Fig. 3.15 A group of Sana'a windows (top left); a close-up photo of three windows showing all the components of the traditional window including the takhrim (top right); takhrim window in the Rock Palace (Dar al-Hajar) at Wadi Dahr, northwest of Sana'a (bottom left); and the traditional window components including the takhrim from inside the room (bottom right) [39]

design variations in terms of size, shape and number, based on climatic influences and functional requirements [38]. They categorized the window design according to its components and described the geometry and function of each component. They concluded that the breakdown of the window into components improves integration of the different functions of the window and provides high flexibility in controlling each function. The four parts of the indigenous window are described below (Fig. 3.1; Figs. 3.14, 3.15 and 3.16):

Fig. 3.16 Windows and takhrims: in a traditional building, Al Hajarayn Wadi Dawan, Yemen (left); in the old customs building in Al-Hudaydah, Yemen (right) [40, 41]

1. *Upper section "qamarya"*: the name "qamarya" comes from qamar in Arabic, meaning moon. Qamarya is a very distinctive multicoloured stained glass window that graces Yemen's buildings and adds a great beauty to Sana'a's distinctive skyline. It is also used widely as a symbol to express Yemeni culture. It sits above the main window and casts its coloured patterns on the room (Figs. 3.14 and 3.15). It has two layers: the inner one is flush with the wall's inner surface of the room, and it has decorative design made of leaves or geometric patterns formed with gypsum frames and coloured glass, and the outer one is flush with the façade surface. It is similar to the inner one in composition, but it is usually made with a different ornament pattern and uncoloured glass. People learned through long experience that making the ornament styling different between the inner and outer layers and using coloured glass in the inner one only give the best option for appealing character. The shape of the outer frame for a wide window is half a circle with the curve positioned upright. When the window's width is narrow, the two end points of the arch are extended vertically underneath to include a rectangular neck. In the more ancient houses, the masonry work of the arch was built to include two full-circle openings that were filled with high-quality semi-transparent alabaster. Only fine types of alabaster were used, ones that had high

ability to diffuse the natural light throughout the indoor space. Qamarya is the only part that involves a material other than wood to frame the stained glass. It is a sealed window (not operable), so it allows only daylight to pass through but not air. Its main function is to provide daylight into the interior space and add vibrancy and aesthetics to the rooms from inside and to the façades from outside, even at night-time. When the shutters of the main window (lower opening) are closed to block direct sun or wind, the qamarya gives comfortable soft lighting with beautiful warmth feeling.

2. *Main window "taqa"*: Taqa is the main window that has a width ranging from 50 to 150 cm. The size is dependent on "which floor"; the traditional applied rule is "the wider window goes for the higher floor." That's because in the lower floors, near the public streets, narrow windows are preferable to maximize protection and privacy, while in the higher floors, large widows are preferable to maximize access to nearby views (e.g. garden of the housing cluster) and/or to far views (e.g. the beautiful mountainous scenery around Sana'a City). Taqa takes the shape of a rectangle or square (Figs. 3.14, 3.15 and 3.16). It could be made as a single layer that includes two shutters of wooden frames holding transparent glass or as a double layer by adding another wooden frame from outside holding wooden shutters. The shutters of both layers are operable. The two glass shutters of the internal layer can be opened by pulling them towards the inside of the room, while the two wooden shutters of the external layer can be opened by pushing them towards the outside of the room. Iron hooks are used to fix the shutters in their opening positions. Taqa is the main provider of air for ventilation, sunlight for warming the rooms, daylight and connection to outside views. The outside shutters can also provide security when they are closed at night or when nobody is in the house. The dynamic mechanism of its components gives high flexibility to control access of sunlight, daylight or ventilation. When all the shutters are closed in the extreme conditions (too cold, too hot or too bright) or at night, it blocks their passage depending on the other parts (the qamarya and shaqus) to provide the minimum needs of air or light.

   In some cases, especially in the houses and palaces of wealthy people another very delightful feature is added to the main window (taqa) from the outer surface. It is called in the Yemeni dialect "takhreem" or "mashrabya" (Fig. 3.1; Figs. 3.15 and 3.16). It is a beautifully handcrafted screen that consists of huge number of tiny pieces of wood members, chosen from the finest wood species, that are fitted into each other to generate intertwining lines or curves that follow an Islamic ornamental pattern. The patterned of engraved wood includes one large operable shutter, or two small ones, that can be opened to the outside. The mashrabya is carried by two cantilevers that are positioned beneath its lower surface, at/near the right and left edges. These two cantilevers are also made of engraved wood that integrates with the pattern of the screen. The mashrabya is positioned so that its top surface is flush with the adjoining lower surface of the kunnah; together they look as one integrated element.

3. *The vent "shaqoos"*: shaqoos is a very small window (around 40 cm in height and 20 cm in width) that takes the shape of a rectangle (Figs. 3.14 and 3.15).

According to the traditional standards, two vents are positioned symmetrically on the right and left sides of the taqa at a high level near the ceiling. The shaqoos is made of a single operable wooden shutter frame holding transparent glass. The shutter can be opened by pulling it towards the inside of the room. An iron hook is used to fix the shutter in its opening position. The main function of the shaqoos is to provide ventilation at a high level of the room near the ceiling. It helps to minimize space overheating by removing the heat that can be accumulated near the ceiling. It works with the taqa (the main window) by promoting passive cooling using stack (or chimney) effect.

4. *The overhang "kunnah"*: kunnah is a horizontal overhang that is made of wood (Figs. 3.15 and 3.16). The width of the kunnah is a bit wider than the main window (taqa). The depth (or projection size from the façade) is relatively small, 40–60 cm. It might seem to many that this size would not be sufficient to cast shadows on the window. Yet, in Sana'a latitude (15.3694°N), a short overhang like the traditional kunnah would work effectively as a shade provider in the summer. That's because the sun position at this latitude during the overheated hours of the day (2–4 p.m.) is very high; in other words, the sun altitude angle is almost perpendicular to the earth surface. The kunnah functions also as a rain protector for the window beneath it. It is usually covered by thick waterproof paints to protect it from damage caused by rain/moisture. Kunnah also adds aesthetic value to the façade's design. Its wood is carved with ornamented edges (sometimes painted with a cheerful colour), which makes it stand out as a vibrant element.

Doors/Gates

The traditional gates in Yemen were made from strong wood like the talh to provide high protection and security. The range of the gate's size and strength varies greatly from the huge city gates like the historical gates of the Old City of Sana'a (e.g. the Gate of Yemen or pronounced in Arabic "Bab Al-Yaman," which is the main gate of Sana's old fortified wall) to the small private homes' entry gates. Gates with intermediate size and strength can be found in public buildings such as mosques, shops and government buildings.

Wood was also used to create doors that make separations between a main hall (or lobby) and individual rooms or bathrooms. Sometimes, doors or some kind of partitions were also used to create separations between zones within the same floor especially where higher privacy level is needed for family members. Fine-quality wood species like the Ethel and Sedr in Shibam, Hadramout, and the Tanab in Sana'a were used to make doors especially when engravings were carved as these types of woods had fewer defects and could withstand the strong works applied by the cutting or chiselling tools [4, 5, 42]. The fancy doors were inlayed by decorative accessories and garnishes made of copper, ivory, pearl or iron. Their craftsmanship and elegance are astonishing (Figs. 3.17 and 3.18).

Fig. 3.17 Double-shutter door in Sana'a (left); single-shutter door in Sana'a (2nd left); door in Ibb (third left); and door in Ibb (last) [43–45]

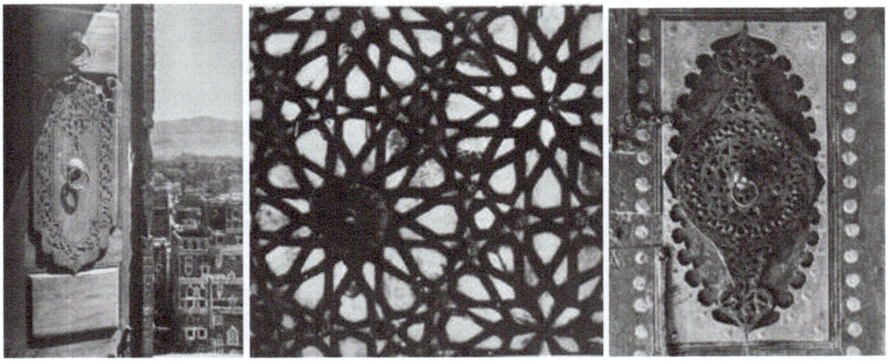

Fig. 3.18 Example of an engraved wooden door, Sana'a (left); door engraving in Al-Motawakel Al-Shahari house (eighteenth century), Old City of Sana'a; engraved wooden door in Talha Mosque (eighteenth century), Old City of Sana'a (right) [46]

Wooden Squared Encasings (Mosandaqat)

The architecture of the Yemeni mosques was given a special attention in the design of their ceilings, minbars (podia) and mihrabs (semicircular niche that indicates the qibla) [47].

The ancient Great Mosque (or Grand Mosque) in Sana'a is located just east of the old Ghumdan Palace site. It is part of the UNESCO World Heritage Site of the Old City of Sana'a. It was built in the seventh century, partly from the materials of the famous Saba'ean Ghumdan Palace. According to the authentic Islamic scriptures, the mosque's history goes back to the period of Muhammad. The building has undergone several renovations in later centuries [48]. An important archaeological find was the Sana'a Manuscript, discovered there during restoration in 1972. It is one of the oldest Quranic manuscripts in existence [49]. The Great Mosque has a splendid wooden ceiling and mihrab (Fig. 3.19). Underneath the structural roof and fixed to it,

Fig. 3.19 Mosandaqat of the ancient Great Mosque, Sana'a, Yemen (left); mosandaqat of the Al-Wahsh Al-Jarani ancient mosque, Wessab Al-Aali village, Dhamar

beautiful ceilings of wooden encasings (mosandaqat) can be found in the main spaces of the mosque. These elements were constructed by wood and shaped in a variety of patterns. Their craftsmanship and elegance are astonishing. The variation in pattern was used as a way to identify or create distinctions between different spaces. This was often combined with varying the level of the patterned surface to indicate different periods of construction. These decorative (non-structural) ceilings were constructed by several rows of purlins that had a square cross section (20 cm thick) extending over the span of the space, crossed by another secondary row of thinner purlins (10–15 cm thick) that are perpendicular in direction. This structure formed coved square spaces between the purlins (e.g. the encasings or mosandaqat). The lower surface of the purlins (main and secondary) and the mosandaqat were decorated by engraved or painted patterns. The whole structure of the ceiling is bound by wooden wrap or belt on which verses of the holy Quran were engraved in koufi font. The engravings might also include quotes by the Prophet or the names of Fatimid rulers.

The ornamentation was made in one of the following styles:

1. Umayyad style: these included ornamentations characterized by motifs related to the Umayyad period and represented as spiral plant branches from which the acanthus leaves, grape leaves and clusters or pine cedar were organized in a symmetrical order with winged elements combined with pearl beads.
2. Abbasid: these ornamentations are characterized by elements closely related to the Abbasid era with motifs that date back to the era of the Al-Ya'far family, especially the Samara III style with decorative elements of arabesque.
3. Fatimid: The third style came with ornamentation elements that appeared in the AH 6/AD twelfth century, such as the stellar dish, egg and arrow ornamentation and playing peat.

Another case study is Al-Wahsh Al-Jarani ancient mosque that has been existing for more than 300 years. It is located in Wessab Al-Aali village in Dhamar Governorate. It is a small countryside mosque that includes only a prayer hall not exceeding 5 × 4 m and a covered pool for ablution. It was built of stone walls and roofed with wood structure and earth covering. The main attractive feature of this mosque is its beautiful ceiling (see Fig. 3.19) that was made with wooden encasings (mosandaqat). They were made by the finest types of wood decorated with colourful carvings and delightful Islamic patterns.

3.5 Threats and Challenges

Threats of Losing Many Historical Sites and Buildings

The War and the Destruction of Treasured Historical Sites

The multifaceted war in Yemen has caused a great deal of suffering and intolerable devastating conditions including death of many people amongst whom are many children, very poor health conditions, collapsed economy, lack of safety and destruction of cultural and historical sites. The air raids by bombs and air missiles and the ground artillery targeted many historical sites in Sana'a, Saa'da, Zabid, Kawkaban and other cities and towns (Fig. 3.20) [50]. It led to killing many civilians including women, the elderly and children and destroyed treasured historical sites. Many of these sites included irreplaceable artefacts made from traditional handcrafted wood such as historical gates, doors, windows and wall or ceiling decorations. The Middle East Monitor [51] reported that the Saudi-led coalition had destroyed 475 tourist archaeological sites and monuments since the start of its aggression in 2015. Also, more than 95% of employees in the tourism industry lost their jobs. According to statistics released by the ministry, over the past 5 years, the coalition destroyed 25 historic cities, 25 mausoleums and 42 archaeological monuments, in addition to 252 hotels, 81 restaurants, 12 festival halls, 38 gardens and parks and eight cafes and cafeterias. Many of these destroyed monuments were very treasured like the

Fig. 3.20 The Republican Palace destroyed by Saudi-led airstrikes, in Sana'a, Yemen, Wednesday, Dec. 6, 2017 (left); Yemenis search for survivors under the rubble of houses in a UNESCO-listed heritage site in the old city of Yemeni capital, Sana'a, following an overnight Saudi airstrike, June 12, 2015 (right) [52, 53]

Republican Palace that was destroyed by the Saudi-led airstrikes, in Sana'a, Yemen, on Wednesday, Dec. 6, 2017. To give some examples, Fig. 3.20 shows two historical sites that were targeted by the airstrikes in Sana'a: The Republican Palace [52] and a destroyed location in the old city of Yemeni capital, Sana'a, the UNESCO-listed heritage site.

Damages Caused by Heavy Rains

The effect of the global warming and climate change resulted in changes in the rainfall patterns. This has led to higher frequency of droughts that impaired hydropower production and caused an increase in floods that significantly raise [54]. Ancient cities of Sana'a, Zabid and Shibam, Hadramout, which are inscribed on the

Fig. 3.21 Workers knock down a rain-damaged building at the UNESCO World Heritage Site in the Old City of Sana'a, Yemen, August 9, 2020 [55]

World Heritage List, are currently exposed to the dangers of torrential rain witnessed in Yemen recently. This constitutes a great threat to them, especially after it caused the demolition and collapse of a number of landmarks and homes of these cities. The heavy rain fragmented the roof cover and reached to the wooden structure that carries the roof and floors that led to the collapse of the entire building or a great deal of it. It also resulted in damaging the external shutters of the windows and the entry doors, all made from traditional wood. Their damage allowed entry of storm water to the interiors causing further damages to the rooms, furniture and the wooden structure of the floors. All these damages including the wooden structures and windows can be noticed in the photos of Fig. 3.21. The GOPHCY has called on the United Nations Educational, Scientific and Cultural Organization (UNESCO) to launch a global appeal to save the ancient cities of Yemen.

The damages caused to the cultural heritage of Yemen as a result of armed conflict and heavy rains were documented in the World Heritage Committee Reports of UNESCO. The damages/threats affected the historic cities of Sana'a, Zabid and Shibam before the last published World Heritage document (WHC/19/43.COM/18) [56, 57], dated March 21, 2019, are outlined in Appendix A and Tables 3.1, 3.2 and 3.3. However, since that date, a lot more war actions and heavy rains that resulted in severe regrettable damages, including total collapses of historic buildings, have happened.

Threats to the Wood Handicraft

War Threats on the Economy and the Traditional Wood Handicraft

The traditional wood handicraft has been one of the oldest and finest heritage in Yemen and the Arab region. The 6 years of devastating war and the collapsed economy, a coproduct of the war and a long period of corrupted governments, have resulted in creating a great danger of incapability to sustain it. The magnitude of this danger is escalating day after day. The danger reached to all aspects of the Yemeni

Table 3.1 Damages and threats affecting the historic city of Sana'a [57]

Damages and threats affecting the historic city of Sana'a	
Threats for which the property was inscribed on the List of World Heritage in Danger	*Damages and threats related to the armed conflict in Yemen*
Factors affecting the property identified in previous reports	• Modern constructions and uncontrolled expansion of commercial activities (issue resolved) • Lack of a safeguarding plan (issue resolved) • Flyover bridge project (issue resolved) • Uncontrolled vertical and horizontal additions • Management activities (use of inappropriate building materials and techniques) • Densification of the historic fabric through occupation of green areas • Functional decay of the residential neighbourhoods • Continuing vulnerability of the property, as a result of extreme conditions since 2011 • *Threats arising from the armed conflict in Yemen* • Physical damage and instability of buildings • Urgent need for shelter for displaced residents • Desertification of green areas and public gardens/orchards • *Problems with the network for the evacuation of rainwater*
Conservation issues presented on March 21, 2019 (summary of main points only)	• "The armed conflict in Yemen continues to threaten the Outstanding Universal Value (OUV) of the property and to cause economic and social impact. The Al-Qassimi, Alfolihi, Madrassa, Al-Bakirya and Bahr Rajraj neighbourhoods remain affected" • There is little available support or funding to carry out maintenance and conservation actions nor to establish and implement corrective measures aimed at removal of the property from the List of World Heritage in Danger. The resulting economic and social pressure has resulted in building violations, which are being addressed through demolition orders and education • The High Committee for Old City of Sana'a Protection has been reactivated to raise awareness, seek funds, encourage community participation and monitoring and safeguard the property. The first national symposium for the preservation of Sana'a heritage took place in July 2018, focusing on sustainable protection and conservation of the old city, on support for the General Organization for the Preservation of Historic Cities in Yemen (GOPHCY) and on building a common vision for strategies needed to prevent the destruction of Yemeni heritage through cooperation between civil society and the government • The State Party requires assistance in ensuring the protection of its heritage, as well as support for institutional and legislative preservation processes

life and not limited to the wood handicraft. This includes also all the expert builders and specialist labourers working in the traditional masonry work, metal work, gypsum work, glass work, etc. Many people have not received their salaries since 2015 due to the collapsed economy and lack of currency. The traditional workforces and the end customer are both hardly hit by this miserable situation.

Table 3.2 Damages and threats affecting the historic city of Zabid [57]

Damages and threats affecting the historic city of Zabid	
Threats for which the property was inscribed on the List of World Heritage in Danger	• Serious deterioration of the built heritage (a high percentage of the residential houses being replaced by concrete and multistorey buildings) • The remaining houses in the city are rapidly deteriorating, due to the prevailing low income of the inhabitants • Since the souq activities have been transferred outside the city, the ancient souq is almost empty and free from any type of activity, and the shops are falling apart • The traditional economic role of the city has vanished • The city in general, is lacking any conservation *and rehabilitation strategies* • *Threats arising from the armed conflict in Yemen*
Factors affecting the property identified in previous reports	• Serious degradation of the city's heritage (many houses and the ancient souq are in an alarming state of deterioration) • Large percentage of the city's houses replaced by inappropriate concrete buildings • Large sections of the city's open spaces have been privatized, either illegally or informally, and more than 30% of these are built up • Reduction in support and resources arising from political and socio-economic disturbances
Conservation issues presented on March 21, 2019 (summary of main points only)	• Armed conflict in close proximity to the property. The State Party has expressed particular concern about the bombardment of areas surrounding the property and groups that might reach the property and cause damage to monuments, despite the efforts of the Ministry of Culture and General Organization for the Preservation of Historic Cities in Yemen (GOPHCY) and ongoing cooperation *with the local council and Zabidi communities* • Owing to the absence of international organizations and the deteriorating economic situation of the country, as well as the unstable security situation, GOPHCY has not been able to take extensive precautionary measures • UNESCO has aided through training programmes for GOPHCY, the General Organization of Antiquities and Museums (GOAM). Urgent financial assistance is still required for the physical conservation of buildings and, thereby, to support local communities and post-conflict recovery, which would eventually contribute to the removal of the property from the List of World Heritage in Danger • GOPHCY still lacks the basic tools for maintenance activities and lacks organizational support • There is an ongoing urgent need for the support of the international community, in addition to the efforts of the World Heritage Centre, Advisory Bodies and existing donors.

Threats Caused by the Destruction of the Power Outages

The ongoing war in Yemen since 2014 has resulted in many severe problems including a miserable economic situation and the destruction of many power stations that supply electricity to various regions. This led to complete electricity outages in most

Table 3.3 Damages and threats affecting the historic city of Shibam, Hadramout [57]

Damages and threats affecting the historic city of Shibam, Hadramout	
Threats for which the property was inscribed on the List of World Heritage in Danger	• Threats from natural elements. • Lack of organizational support and material resources for conservation. • *Threats related to the armed conflict*
Factors affecting the property identified in previous reports	• Floods (issue previously reported as being resolved) • Poor maintenance (issue previously reported as being resolved) • Damage to historic buildings • Reduction in support and resources arising from political and socio-economic disturbances • Armed conflict situation since 2015 • *Threats from rain and floods*
Conservation issues presented on March 21, 2019 (summary of main points only)	• Despite the deteriorated condition of the property, economic and social constraints and a lack of external support, the State Party continues intensive efforts to preserve the cultural heritage values of the property. GOPHCY has acted quickly within its available resources to stabilize structures affected by floods and armed conflict. However, basic maintenance tools and financial and organizational support remain lacking. The State Party has also continued to consult with the World Heritage Centre and Advisory Bodies • The report "Conservation Status of Shibam Hadramout 2018–2019, Strategy for the Management of the Historic City of Shibam" has been prepared to provide a management strategy for the property, with regard to its physical condition, and the political and conflict context in Yemen. GOPHCY has developed a series of plans and programmes, including engagement with local authorities and communities, improved communication with international organizations, implementation of building restoration programmes and workshops and awareness programmes, bulletins and urgent appeals for rescue and restoration • *Major projects undertaken include the preparation of a plan for drainage inventory, studies for the repair of flood and conflict-damaged buildings, a study for the restoration and maintenance of the historical palace of Sayoun and the preparation of a study for a proposed new government complex*

regions, or continuously recurring in other ones. This miserable situation forced the carpentry workshops to go back to their primitive traditional tools such as the hand-saw or chisel to cut and engrave the wood. Yet, these old tools cannot be very productive unless used with high-quality wood such as "Tanab." Tanab was highly known as the best wood for products that require fine detailing and applying carving and engraving work. That is because it had many special characteristics: it was durable and rarely had natural defects (such as knots or bad grains), it could withstand well against the applied processes of carving and chiselling without losing its edges or causing defects and it looked very elegant when crafted carefully. After Yemen was hit by bad economic situation, Tanab wood became very rare and expensive, and hence, most people cannot afford to buy it.

Threats Caused by the Imported Cheap Wood

Bad economy resulted in higher dependence on cheap kinds of imported wood replacing the traditional fine-quality species. The high dependence on the non-local wood resulted in creating many challenges for the Yemeni traditional carpenter and wood carver and the wood market in general. Cheap kinds of imported wood are filling up the market now because of its much lower prices compared to the local wood. It is imported from cheap kinds to lower the prices so that the end customer can afford to buy it. Because it is cheap quality, the carpenter and wood carver face many difficulties in producing a high-quality work as they used to make before.

Threats Caused by the Ready-Made Products

The traditional wood handicraft has been affected hardly by the ready-made imported wooden products. Ready-made wooden products especially doors are available and common in the Yemeni market today. These doors are imported from many countries; the most common ones are imported from China, Malaysia and Turkey. In today's world, almost everything can be obtained by mail orders. This includes wooden products that are used in constructing or furnishing buildings. For instance, people buy all sorts of ready-made doors through mail orders as they find this method easy and attractive as it gives them possibility to see and choose from many options. Although the war situation and the forced siege on airports and marine ports make the mail ordering method difficult to depend on, people in certain areas in Yemen still find it possible and perhaps attractive. That's because the authorities left some of the border entries from Saudi Arabia and Oman open yet under highly watchful conditions.

Threats Caused by Change of People's Taste and Availability of Ready-to-Use Templates

People's taste of art and appreciation of the local style have changed over the years. One can see today ornamented wood doors, windows, furniture, walls and ceiling with odd engravings, and most influences came from the Arab Gulf states and other Asian cultures especially from India, Iran, Malaysia and China. Sometimes, they are even mixed together in one artefact or in multiple ones in the same room. Such new habits started by the people who went to and stayed in the Arab Gulf states for jobs and came back to Yemen. What made the problem worse is the wide availability of easy-to-use templates that can be downloaded from the Internet and the use of new technologies such as the CAD/CAM software and CNC machines, as these opened the door to unlimited choices of new patterns.

3.6 Can We Solve the Problem?

"Can we solve the problem" is a very tough question to raise amidst all the tragedies happening in Yemen today: raging war, collapsed economy and miserable health conditions including cholera and COVID-19 breakouts. The Yemeni people cannot solve all these tragedies alone, especially that external parties (regional and maybe international) are involved in keeping the war raging by their flowing finances to fund and train militias and supply them with all necessary weapons and intelligence. It seems also that the international community does not pay much attention to what is happening in Yemen. In spite of this, hope must be embraced. Since the wood handicraft cannot be separated from architecture as a whole, a general preliminary framework for sustaining the historical sites is proposed below. It might serve as a launch to build up on it and create a comprehensive plan when the international community manages to stop the war.

General Proposed Framework for Sustainability

- *Campaigning actions to call for help and engage world organizations*:
  - Engage world multilateral aid organizations such as the United Nations World Heritage Committee.
  - Engage international organizations (only the countries that are not involved in the armed conflict).
  - Engage research centres and academic institutions.
- *Create a financial system to support managing, restoring and sustaining the historical sites*:
  - Establish fund-raising programmes.
  - Establish a dedicated financial mechanism or fund.
  - Allocate money for (1) reconstructing and restoring, (2) monitoring the properties, (3) education and training, (4) research and publications (5) travel funds and conference organization, (6) campaigning and fund-raising, (7) rewarding cooperative inhabitants and penalizing non-cooperative ones, (8) initiate sponsorship programmes that provide suitable income and living conditions for the master builders and craftsmen and (9) loans to help inhabitants in the restoration processes.
- *Technical actions to safeguard the historical sites*:
  - Reconstruct the properties (buildings, streets, city walls, water channels, etc.) that collapsed due to the armed conflict and heavy rains.
  - Repair all the urgent damages (ruins happened to roofs, walls, windows, doors, etc.) caused by the armed conflict and heavy rains.

- *Educational actions to sustain the handicraft*:
  - Revitalize the traditional way of life by creating centres of training for young workers while there are a few remaining masters of old building techniques still alive.
  - Keep track of the masters of old building techniques (traditional master builders and craftsmen), and create a database of their names, addresses, specialties (carpentry, wood carving/engraving, etc.), architecture style (Sana'ani, Shibami, Zabidi, etc.).
  - Initiate sponsorship programmes for the master builders and craftsmen.
  - Establish technical school and research centre in every historic city in which the traditional masters of buildings train students and younger craftsmen.
  - Engage architectural schools in Yemen and around the world in collaboration research and training programmes with the local party (GOPHCY).
- *Legal actions*:
  - Introduce a system to reward best practices made by inhabitants.
  - Introduce an efficient and speedy system to detect the violations made by inhabitants or the public and give warnings.
  - Design a scaled system of warnings with a grace period after which fines can be charged.

3.7 Epilogue

Yemen has many treasured historical sites that include remarkable cities, towns and buildings. Many examples still stand up today proving the long history of Yemen and its rich cultural and architectural heritage. The richness of the Yemeni heritage is a result of several factors: (1) multiple layers of interwoven historical influences from different cultures and periods (Arab, Aksumite, Persian, Ottoman and Indian), (2) different ethnic religion transformations (including Judaism, Christianity, Islam) throughout the history and (3) great diversity of geography (variations of landforms and climates and richness of natural resources for building materials). These all together created a variety of unique regional architectures filled with splendid artefacts and handmade woodworking.

The chapter presented the traditional methods of using wood as (1) a structural system to support the roof, the intermediate floors, the traditional sanitary duct system, the staircase and the bridge connecting two buildings; (2) a reinforcing material of buildings' mud walls and cities defensive walls; (3) an environmental control system for wind, sun, daylight and view using different kinds of windows and window components; (4) an aesthetic material that created splendid carved and engraved artefacts; (5) a protection provider to cities and building entries using heavy fortified gates; and (6) a space divider and privacy provider using doors.

It presented a discussion of the historical and current situation in terms of the technical aspects of the traditional woodworking and provided many architectural case studies to illustrate the ideas. The difficulties to travel and collect onsite information due to the ongoing war and enforced siege on the country's entries created great restrictions, which obliged the author to depend on literature review and communication through social media and phone calls with experts and specialists.

The chapter discussed at the end the tragedies imposed by the ongoing war and collapsed economy that threaten the whole country and people today and how these struck the Yemeni vernacular architecture and traditional wood handicraft in particular. The ongoing war in Yemen is multifaceted with external parties (regional and international) providing the leadership, funding and weaponry that create and train the militias and feed the military confrontations. This is accompanied by the outbreak of COVID-19, the reduction in humanitarian operations and the volatility of the currency that all together have led to a deterioration in socio-economic conditions:

When this chapter was written, no light can be seen at the end of the tunnel. The Yemeni people cannot solve these tragedies alone. The international community, especially powerful politicians, seem to not be much concerned about the deaths of the Yemenis (mostly civilians and children) or the destruction of their culture. Until the day when the international conscience awakens to the Yemen's catastrophe, very little can be done by a poor country like Yemen; and hence it would be immature to propose comprehensive technical solutions at this point of time. Any solution should start first by stopping the main cause (i.e. the war) and saving the people's lives. Nevertheless, a general preliminary framework was proposed that might serve someday as a starting point to build up on it.

Acknowledgements Special thanks go to artist Iman Al-Sallal who created all the nice illustrations in this chapter.

Appendix A: UNESCO World Heritage Committee Report

The World Heritage document (WHC/19/43.COM/18) of the World Heritage Committee (WHC), 43rd session, concerns the protection of the world cultural and natural heritage. The convention that took place in Baku, Republic of Azerbaijan, on June 30 to July 10, 2019, included the threats to the historic cities of Sana a, Shibam, Zabid and Socotra [56]. A summary is given below.

The WHC:

- Expressed its continuing concern at the damage caused to the cultural heritage of Yemen as a result of armed conflict and that the old cities have incurred irreversible destruction and continues to be vulnerable, owing to the current security situation, ongoing social change and continuing lack of support and resources for both heritage management and physical conservation.

- Reiterated the need for a joint World Heritage Centre/International Council on Monuments and Sites (ICOMOS) Reactive Monitoring mission to advise on repair and conservation works and to contribute to the development of a set of corrective measures and a timeframe for their implementation, as well as the desired state of conservation for the removal of the property from the List of World Heritage in Danger (DSOCR), as soon as the security situation in Yemen has improved.
- Urged all parties involved in the conflict to refrain from any further action that would cause damage to the cultural heritage and the outstanding universal value (OUV) of the property and to fulfil their obligations under international law by taking all possible measures to protect such heritage, in particular the safeguarding of properties on the World Heritage List and those included in the Tentative List of Yemen, and also encouraged all concerned stakeholders to unite for the preservation of cultural heritage in Yemen.
- Reiterated its previous calls to the international community to provide technical and financial support, including through the UNESCO Heritage Emergency Fund, for the implementation of the Emergency Action Plan for the Safeguarding of Yemen's Cultural Heritage, adopted at the UNESCO expert meeting in July 2015, including funding for capacity building and first aid restoration and protection measures, and calls on the World Heritage Centre and the Advisory Bodies to continue providing technical assistance and support where needed.

The damages/threats that affected the historic cities of Sana'a, Zabid and Shibam before the World Heritage document (WHC/19/43.COM/18), dated March 21, 2019, are outlined here in Tables 3.1, 3.2 and 3.3 [57]. Since that date, much higher number of damages has been witnessed and presented by the world news especially the ones caused by the armed conflict/airstrikes and heavy rains.

References

1. Residential Structural Design Guide. (2017, October). *A state-of-the-art engineering resource for light-frame homes, apartments, and townhouses* (2nd ed.). Prepared for U.S. Department of Housing and Urban Development, Office of Policy Development and Research, Prepared by Coulbourne Consulting.
2. Encyclopaedia of Yemen. (1992). 2nd, Ahmed Jaber Afif, Alafif Cultural Foundation.
3. Wikipedia Arabic, Sedr. (2020, July 23). Retrieved from https://ar.wikipedia.org/wiki/%D8%B3%D8%AF%D8%B1
4. Know your country: the carpentry profession and Yamani's ability to twist inanimate objects and turn them into a vibrant art. (2016, December 21). Retrieved from https://youtu.be/0MxPCnpYVF8
5. Know Your Country—Report: Wood carving an ancient Dhamari art. (2017, May 12). Retrieved from https://youtu.be/Unf-cL3OCLQ
6. Hunker. (2020). What is Meranti wood? Retrieved from https://www.hunker.com/13401431/what-is-meranti-wood
7. WoodMagazine.Com, Meranti. (2020). Retrieved from https://www.woodmagazine.com/materials-guide/lumber/wood-species-2/meranti

8. Wikipedia Beech. (2020, September 6). Retrieved from https://en.wikipedia.org/wiki/Beech
9. Wikipedia, Teak. (2020, September 13). Retrieved from https://en.wikipedia.org/wiki/Teak#cite_note-GRIN-2
10. The Wood Database, Teak. (2020). Retrieved from https://www.wood-database.com/teak/
11. Wikipedia, Arabia Felix. (2020, September 2). Retrieved from https://en.wikipedia.org/wiki/Arabia_Felix#cite_note-4
12. Webster's New Geographical Dictionary. (1972). Springfield, MA: Merriam-Webster.
13. Graf, D., Talbert, R., Gillies, S., Elliott, T., & Becker, J. (2014). Places: 746710 (Arabia Eudaemon). Pleiades. Retrieved November 1, 2014.
14. Sergeant, R. B., & Lewcock, R. (1983). San'a: An Arabian Islamic City, London.
15. Reich, B. (1990). *Political leaders of the contemporary Middle East and North Africa: A biographical dictionary*. Greenwood Publishing Group.
16. Dark Roasted Blend, The Most Alien-Looking Place on Earth by Abrams, Avi. (2008, September 4). Retrieved from http://www.darkroastedblend.com/2008/09/most-alien-looking-place-on-earth.html
17. Huntingford, G. W. B. (1980). *The Periplus of the Erythraean Sea*. Hakluyt Society.
18. SABA Net News, EU to protect Socotra archipelago environment. Saba Net. Yemen News Agency, SABA. (2008, April 15). Retrieved from https://www.saba.ye/en/news151852.htm
19. Scholte, P., & De Geest, P. (2010). The climate of Socotra Island (Yemen): A first-time assessment of the timing of the monsoon wind reversal and its influence on precipitation and vegetation patterns. *Journal of Arid Environments, 74*(11), 1507–1515.
20. Flickr, Jacgroumo – Yemen Lakamat. (2008). Retrieved from October 25, 2008, from https://www.flickr.com/photos/143032208@N03/28606748423/in/photostream/
21. Wikimedia Commons. (2019). *A photographic journey through Southern Yemen, Wadi Dowan, Hardrmout, Matthew Reichel*. Retrieved from December 14, 2019, from https://commons.wikimedia.org/wiki/File:Shibam_Wadi_Hadhramaut_Yemen.jpg
22. Al-Sallal, K. A., Ayssa, A. Z., & Al-Sabahi H. A. (1995). Thermal performance and energy analysis for Sana'a vernacular house. In S. Ben Ghadi (Ed.), *Proceedings of applications on renewable energy in Yemen workshop*. Yemen: Aden University.
23. Al-Sallal, K. A. (1996a). Traditional methods in new forms: insight to achieve energy conservation in the modern Sana'a house. In *Proceedings of world renewable energy congress IV in Denver*.
24. Al-Sallal, K. A. (1996b). Solar access/shading and building form: geometrical study of the traditional housing cluster in Sana'a. In *Proceedings of world renewable energy congress IV in Denver*.
25. Al-Sallal, K. A. (2001). The balanced synthesis of form and space in the vernacular house of Sanaa: bioclimatic and functional analysis. *Architectural Science Review 44*(4), 419–428. https://doi.org/10.1080/00038628.2001.9696922.
26. Al-Sallal, K. A. (2004). Sana'a: transformation of the old city and the impacts of the modern era. In Y. Elsheshtawy (Ed.), *Blurring identities: the middle eastern city in an age of globalization*. London: Routledge.
27. Al-Sallal, K. A. (2013). Vernacular tower architecture of Sana'a: Theory and method for deriving sustainable design guidelines. In A. Sayigh (Ed.), *Sustainability, energy and architecture: Case studies in realizing green buildings* (p. 257). London: Academic Press.
28. Al-Sallal, K. A. (2016a). Passive and low energy cooling. In *Low energy low carbon architecture* (pp. 57–102). Boca Raton: CRC.
29. Al-Sallal, K. A., & Rahmani, M. (2019). Vernacular architecture in the MENA region: Review of bioclimatic strategies and analysis of case studies. In *Sustainable vernacular architecture* (pp. 23–53). Cham: Springer.
30. Google Earth Pro. (2018). https://earth.google.com/web
31. 26-SDNS. (2011). 26 September Daily News Site. The Directorate of the old Sana'a is going out of its way on Friday in support of legitimacy, Friday April 1st, 2011. Retrieved April 16, 2018, from http://www.26sep.net/news_details.php?sid=72682

32. Bijil, A. (2015). In Yemen, there's a city full of 500-year-old skyscrapers made of mud. CityMetric (now City Monitor). Retrieved October 12, 2015, from https://www.citymetric.com/skylines/yemen-theres-city-full-500-year-old-skyscrapers- made-mud-1462
33. Helfritz, H. (1937). Land without shade. *Journal of the Royal Central Asian Society, 24*(2), 201–216. https://doi.org/10.1080/03068373708730789.
34. Dethier, J. (2020). Inhabiting the earth: A new history of raw earth architecture. *The Architectural Review*. Retrieved January 31, 2020, from https://www.architectural-review.com/essays/inhabiting-the-earth-a-new-history-of-raw-earth-architecture
35. Maréchaux, P., & Maréchaux, M. (1997). *Impressions of Yemen*. Flammarion.
36. Meadowcroft, H. (2018). Mud Brick Architecture of Yemen, By Howard Meadowcroft. *Global dispatches*. Retrieved September 2, 2018, from http://www.theglobaldispatches.com/articles/mud-brick-architecture-of-yemen
37. Lewcock, R. (1986). *The Old Walled City of Sana'a*. Paris: UNESCO.
38. Al-Sallal, K. A., & Cook, J. (1992). Sana'a historical windows: integration between comfort and aesthetics. In *Proceedings of the National Passive Solar Conference* (Vol. 17, p. 197). Boulder, CO: American Solar Energy Society.
39. Smugmug, P. S., & Palarczyk, M. (2020). *Yemen, an Arabian fairy-tale*. Retrieved from https://paulsmit.smugmug.com/Features/Asia/Yemen/
40. Inertia Network, The Inertia Guide to South Yemen, Matthew Reichel. (2020, April 1). Retrieved from https://travel.inertianetwork.com/inertia-content/2020/4/1/the-inertia-guide-to-south-yemen
41. Flickr, Jacgroumo – Yemen Al-Hudaydah Old Customs. (2008). Retrieved November 27, 2008, from https://www.flickr.com/photos/143032208@N03/29119102052/in/photostream/
42. A Yemeni sculptor specializing in Shabami Hadrami woodcarving. (2018, October 24). Retrieved from https://youtu.be/c4U0D4l4gPw
43. Pinterest. (2020). *Old door, Sana'a*. Retrieved from https://www.pinterest.co.uk/pin/239394536415320224/
44. Wikimedia Commons. (2020). *Sana'a Door, Yemen*. Retrieved from https://commons.wikimedia.org/wiki/File:Sanaa_Door,_Yemen_(10268040125).jpg
45. Flickr, Jacgroumo – Door of a house that belonged to a Jewish craftsman, Ibb. (2008). Retrieved October 23, 2008, from https://www.flickr.com/photos/143032208@N03/28606528843
46. Institut du monde arabe (Paris), Maréchaux, P., Ṣāliḥ, H., & Flèche, I. (1987). *Sanaa: parcours d'une cité d'Arabie*. Institut du monde arabe.
47. Sana'a, The Foundations of Architectural Design and Urban Planning. (2005). *Organization of Islamic capitals and cities* (p. 312).
48. Talgam, R. (2004). *The Stylistic Origins of Umayyad Sculpture and Architectural Decoration: Text* (p. 112). Otto Harrassowitz Verlag. ISBN 978-3-447-04738-8. Retrieved December 25, 2012.
49. Sadeghi, B., & Goudarzi, M. (2012). *Ṣan'ā' 1 and the Origins of the Qur'ān. Der Islam* (Vol. 87, pp. 1–129). Berlin: De Gruyter. https://doi.org/10.1515/islam-2011-0025.
50. Statista, Number of air raids on Yemen by Saudi Arabian-led coalition from March 2015 to March 2018. (2020). Retrieved from https://www.statista.com/statistics/940899/yemen-air- raids-by-saudi-led-coalition-per-month/
51. MEMO. (2020). Retrieved April 2, 2020, from https://www.middleeastmonitor.com/20200402-yemens-houthis-saudi-coalition-destroyed-475-archaeological-sites/
52. The Washington Times – Friday. (2020). Retrieved December 8, 2017, from https://www.washingtontimes.com/multimedia/image/yemen_66237jpg-e60e1jpg/
53. Press TV. (2020). Retrieved November 21, 2015, from https://www.presstv.com/detail/2015/11/21/438556/yemen-historic-sites-saudi-arabia
54. Mohamed, H. (2017). Yemen between the Impact of the Climate Change and the Ongoing Saudi Yemen War: A Real Tragedy. In *An Analytical Report Published by the Centre for Governance and Peace building Yemen, in Collaboration with Centre for International Development Issues*. Nijmegen: GPBC and CIDIN, Radboud University, The Netherlands.

55. Reuters, World News. (2020). Retrieved August 10, 2020, from https://www.reuters.com/article/us-yemen-floods-sanaa-old-city-idUSKCN2560MK
56. UNESCO, WHC/19/43.COM/18. (2019). Retrieved from https://whc.unesco.org/archive/2019/whc19-43com-18-en.pdf
57. UNESCO, WHC/19/43.COM/7A.Add.2. (2019). Retrieved from https://whc.unesco.org/archive/2019/whc19-43com-7AAdd2-en.pdf

Chapter 4
Timber as a Sustainable Building Material from Old to Contemporary Experiences: Review and Assessment of Global and Egypt's Examples

Mohsen Aboulnaga and Maryam Elsharkawy

The timber roof of a BurJuman shopping mall, Dubai, the UAE (*Source photo credit: Mohsen Aboulnaga*).

> *Good timber does not grow with ease; the stronger the wind, the stronger the trees.*
>
> ***J. Willard Marriott*** *Source: https://www.azquotes.com/quotes/topics/timber.html*

M. Aboulnaga (✉) · M. Elsharkawy
Department of Architecture, Faculty of Engineering, Cairo University, Cairo, Egypt
e-mail: maboulnaga@eng.cu.edu.eg

A. Sayigh (ed.), *The Importance of Wood and Timber in Sustainable Buildings*, Innovative Renewable Energy, https://doi.org/10.1007/978-3-030-71700-1_4

4.1 Introduction

Wood is not only a building material—whether in the skeleton of buildings or in the interior furniture or an insulated material—but also a sustainable material. In various old civilizations such as ancient Egypt, Greece, and Rome, wood (timber) was widely exploited in constructing temples and buildings [1]. Lumber (chopped wood or timber) is contemplated as second in the ladder of building materials after stone [2]. Figure 4.1 presents a sailing boat in the ancient Egyptian civilization where wood and timber were used in the boat and accessories, and Fig. 4.2 also illustrates some examples of ancient Egyptian, Greek, Roman, and Romanesque as well as Amawyi era where wood, stones, and bricks were exploited in their buildings.

Since the industrial revolution and the emergence of iron, steel and cement as well as glass lead to the heavy use of nonorganic materials in the construction sector worldwide [3–5]. Such a trend made it not viable and impossible for prudent architects to consider the utilization of organic material such as wood and timber in any significant, meaningful, and sustainable means and fashion.

It is imperative to distinguish between the wood, timber, and lumber in order to understand the context. The term "wood" is used to describe the substance that makes up the tree; it is the hard, fibrous structural tissue, which is usually found in the trees' stems and roots [6]. However, in North America—Canada and the United States, the word "timber" refers to wood from felled trees. Nonetheless, lumber utters products from wood or timber [7]. Box 4.1 illustrates the difference in brief.

Box 4.1 Difference between wood, timber, and lumber

Wood is the hard, fibrous structural tissue of the tree, which is commonly found in the trees' stems and roots
Wood is often refer to as timber in North America. Lumber refers to wood/ timber products
(Source: https://www.designigbuildings.co.uk/wiki/Timber_vs_wood*)*

Fig. 4.1 Wooden sailing boat of ancient Egypt at the Maritime Museum in Liverpool, the United Kingdom (*Source:* https://www.pinterset.com)

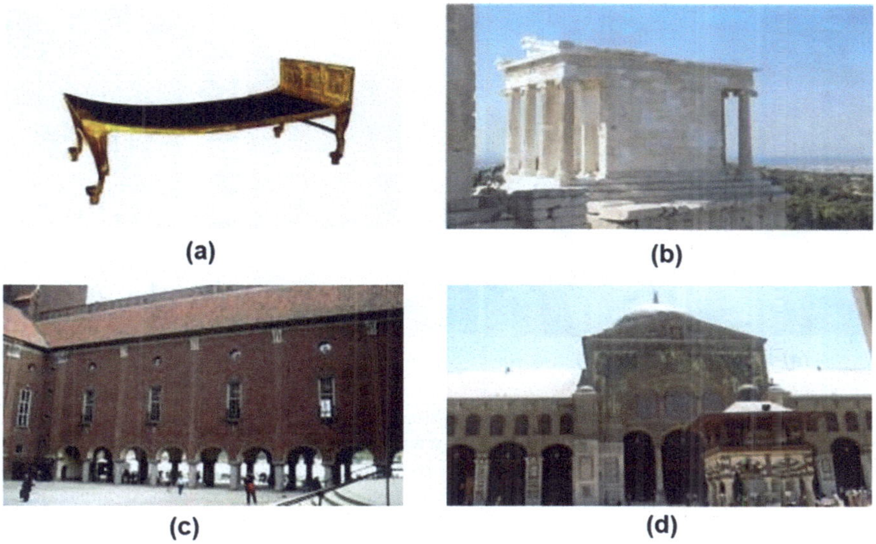

Fig. 4.2 Wood is used in various civilizations. (**a**) A wooded bed of Tutankhamun ancient Egypt (*Source:* https://Tutankhamun-london.com/see-do/). (**b**) Stone used in temple of Athena Nike in Greek civilization, Greece (*Source:* https://alchtron.com/Temple-of-Athena-Nike). (**c**) Wood used in windows of a Romanesque building in the Stockholm City Council, Sweden (*Source/photo credit: Mohsen Aboulnaga*). (**d**) Wood used in courtyard and allies of the Amawyi in Damascus, Syria (*Source/photo credit: Mohsen Aboulnaga*)

Thus, it is considered an environmental-friendly building material. Lumber as a highly sustainable building material has been used in many countries including Europe, Asia, and North America. It's high ability to store carbon content (250 kg/m^3) and release low carbon emissions (15 kg/m^3) has granted it superior sustainable features over steel, concrete, and aluminum that do not store any carbon but release high carbon emissions with 5320, 120, and 22,000 kg/m^3, respectively [8]. Box 4.2 and Fig. 4.3 present these facts. In reality, lumber is obtained from chopped wood (timber) which are naturally growing trees as shown in Fig. 4.4.

Box 4.2 Facts about wood as a sustainable material

Wood has high ability to store carbon content of 250 kg/m^3 and release low carbon emissions 15 kg/m^3

Wood has a superior sustainable features over steel, concrete, and aluminium since the latter constriction material release high carbon emissions at 5320, 120 and 22,000 kg/m^3, respectively

(Source: Braulio-Gonzalo, M, et al. 2014)

Fig. 4.3 Wood is used in various civilizations (*Source of images:* (**a**) *and* (**c**) www.freeimages.com*;* (**b**) *photo credit: Krisztian Tabori; and* (**d**) www.unsplash.com). (**a**) *Trees in Incheon, South Korea* (Photo credit: Mohsen Aboulnaga). (**b**) *Trees in Incheon, South Korea* (Photo credit: Mohsen Aboulnaga). (**c**) *Growing trees in Nanjing, China* (Photo credit: Mohsen Aboulnaga). (**d**) *Beauty of trees in Nanjing, China* (Photo credit: Mohsen Aboulnaga)

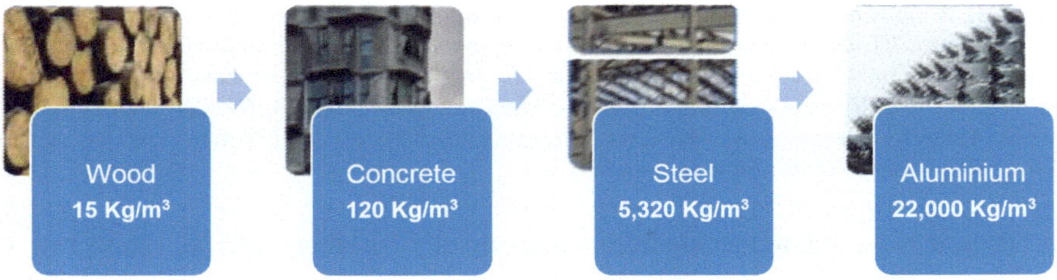

Fig. 4.4 *Wood (timber) is a natural and sustainable material* (Source: lead author)

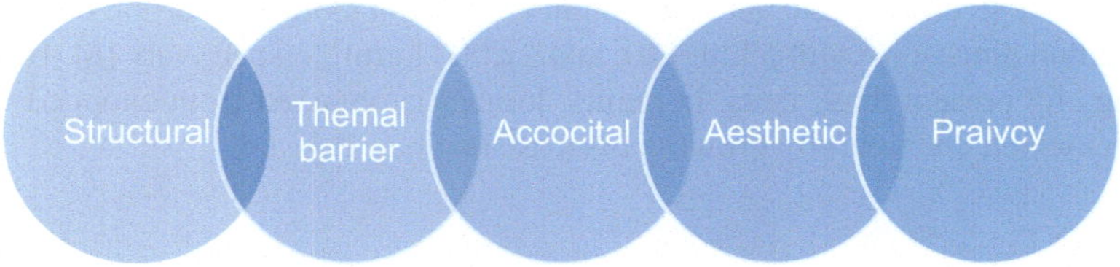

Fig. 4.5 *Timber's various functions* (Source: developed by authors)

Outlook of buildings using timber in Europe, Asia, and North America gives a broad overview of the material usability in buildings that could be integrated in construction for various purposes. Timber, as a building material, is found suitable for usage in buildings' structural elements, façade components, and interior features. Timber is also integrated and highlighted in previous buildings for its diverse roles. Primarily, structural role of timber could be recognized as the most sensitive and delicate structural building material for its restrained nature compared to steel and concrete [9]. Wood also acts in many functions in buildings as illustrated in Fig. 4.5.

Fig. 4.6 The Notre-Dame de Paris cathedral in Pairs, France (*Photo credit: Mohsen Aboulnaga*)

Successful notable timber structures stand today to emphasize the effectiveness of the material in the building construction through ages as revealed in the built skeleton of Notre-Dame de Paris in Paris, France [10]; the Market Square buildings in Brussels, Belgium [11, 12]; Royal Garden in Vienna, Austria [13]; the dome of the Duomo in Florence [14]; the Royal Dusit Palace (summer residence) in the outskirts of Bangkok, Thailand [15]; the old royal building in Seoul, South Korea [16]; and the Jiming Temple in Nanjing, China [17], where it is totally built by timber and shingles. These buildings are illustrated in Figs. 4.6, 4.7, 4.8, 4.9, 4.10, 4.11, and 4.12, respectively.

Most of these structures are constructed back in the nineteenth century, and till date, such examples are proven to be structurally stable, exhibiting the robustness of timber as a structural element. Also, timber in the form of lumber is acting as a thermal barrier in addition to an aesthetic function especially in buildings' façades. This function has been studied to convey the high ability of timber in maintaining low thermal conductivity of exterior extreme conditions to indoor-built spaces [18]; besides, it has been illustrated in building architecture of façade elements, where additional movable and fixed façade shades are added to provide the required thermal barrier to extreme cold and hot climate conditions.

It is imperative to state that timber is also found to be unique, and added aesthetic spirit to the building is expressed in the wooden screen in Nazarbayev University in Astana, Kazakhstan [19]; also the windows' shutters in Rome, Italy [20]; in Pordenone; and in Bolzano, South Tyrol, north of Italy [21]; façades of old buildings in Dan Hugae, the Netherlands; the royal building wooden shutters of Schönbrunn Palace in Vienna, Austria [22]; and a wooden façade of a residential building in Barcelona, Spain [23], as well as window shutters in Abdeen Palace in

Fig. 4.7 The Market Square in Brussels, Belgium (*Photo credit: Mohsen Aboulnaga*)

Fig. 4.8 Royal Garden in Schönbrunn Palace in Vienna, Austria (*Photo credit: Mohsen Aboulnaga*)

Cairo, Egypt [24]. Figures 4.13, 4.14, 4.15, 4.16, 4.17, 4.18, 4.19, and 4.20 present these buildings' examples. Moreover, the carved wooden elements are integrated in old buildings' façades in Casablanca, Morocco [25], and the doors in old building in Sidi Bou Said in Tunisia [26] as well as timber doors utilized in a synagogue in Casablanca, Morocco, are illustrated in Figs. 4.21, 4.22, and 4.23, respectively.

Fig. 4.9 The dome of the Duomo—iconic cathedral in Florence, Italy (*Photo credit: Mohsen Aboulnaga*)

Fig. 4.10 The Royal Dusit Palace (summer residence), the outskirts of Bangkok, Thailand (*Photo credit: Mohsen Aboulnaga*)

Acoustical Role of Timber

The acoustical role of timber in buildings' interiors is widely acknowledged throughout history [27]. This is usually represented in the interior spaces concerned with noise reduction such as inside the NOI center in Bolzano, Italy (Fig. 4.24), and the Great Library of Alexandria, Egypt, is shown in Fig. 4.25. Then, mixed role of timber reveals the significance of timber in multi-functioning in the built environment [27, 28, 29]. Most buildings that depend on timber, as an integrated built material, consider one or more design aspects, for instance, both structural and decorative role in the hanging church (Al-Maghara Church) in Cairo, Egypt [30], shown in Fig. 4.26 and Abbassiya Church well as in Figs. 4.27 and 4.28; in both structures, the acoustical role in the interior of these buildings is clearly manifested.

Fig. 4.11 Wooden gate of an old royal palace in Seoul, South Korea (*Photo credit: Mohsen Aboulnaga*)

Fig. 4.12 Wooden structure of the Jiming Temple in Nanjing, China (*Photo credit: Mohsen Aboulnaga*)

Fig. 4.13 Windows' screens at Nazarbayev University in Astana, Kazakhstan (*Photo credit: Mohsen Aboulnaga*)

Fig. 4.14 Windows' shutters in Rome's building, Italy (*Photo credit: Mohsen Aboulnaga*)

Fig. 4.15 Windows' wooden shutters in Bolzano, north of Italy (*Photo credit: Mohsen Aboulnaga*)

Lessons and Features Learned from the Literature

Based on the literature, lessons revealed the imperfections of timber as a building material as well as the significant role of timber in the construction industry. It was found that timber as a natural building material encounters decaying properties and is considered a major disadvantage of material usage [31]. The rate of timber decay varies in buildings located in different climatic conditions [32]. Also, timber is considered a highly flammable building material [33], and fire safety should be deeply studied and considered in a high-timber-content building type. Though fire accidents in buildings have been well known in old timber structures, nowadays, timber treatments create flame-protective timber elements [34]. Nevertheless, the timber assessment as a sustainable building material should be based on multi-criteria evaluation [35], where the material life cycle should be involved in a cradle to grave

Fig. 4.16 Wooden windows' shutters in Pordenone, Italy (*Photo credit: Mohsen Aboulnaga*)

Fig. 4.17 Old buildings' façades with wooded windows in Dan Hugae, the Netherlands (*Photo credit: Mohsen Aboulnaga*)

Fig. 4.18 An annex building of Schönbrunn Palace in Vienna, Austria showing the shutters' windows. (*Photo credit: Mohsen Aboulnaga*)

Fig. 4.19 Wooden façades of a residential building in Barcelona, Spain (*Photo credit: Mohsen Aboulnaga*)

evaluation method. Timber reuse, recycle, recovery, and landfill analysis specify the sustainability degree of the material usage in buildings [36].

Why Timber Buildings Are Vital in Mitigating Climate Change Impacts?

Buildings built in the sixteenth and seventeenth centuries with timber and stones in cities like those in Florence, Italy, are still lasting till date. These excellent models are low-carbon buildings and contribute to mitigating climate change. Figures 4.29,

Fig. 4.20 One of Abdeen Palace's buildings, Cairo, Egypt (*Photo credit: Mohsen Aboulnaga*)

4.30, and 4.31 present such examples. As a matter of fact, research studies confronted conflicting opinions in regard to the ability of timber as a sustainable material to mitigate climate change impacts [37]. Focusing on both the carbon content in the material and the carbon emissions produced by the material, timber is capable to draw attention in the construction industry in terms of dramatic emissions' reduction. Also, to ensure high material sustainability, certified forests' wood is considered a way to overcome the drawbacks of deforestation and loss of forests in favor of collecting sustainable wood for construction [38], not to ignore the fact that timber is biodegradable and timber wastes are environmentally friendly.

4.2 Objectives

This work is intended to make a significant platform for educating architects to look at subjects normally ignored. The use of nonorganic materials such as cement and iron made it impossible for sensible architects to consider the exploitation of organic material such as wood and timber in any meaningful way. The research work aims at enriching the knowledge base and highlighting the advantages and disadvantages

Fig. 4.21 Wooden façades in old buildings in Casablanca, Morocco (*Photo credit: Mohsen Aboulnaga*)

Fig. 4.22 Timber used in inside and entrance of a synagogue in Casablanca, Morocco (*Photo credit: Mohsen Aboulnaga*)

Fig. 4.23 An old building in Sidi Bou Said in the city of Tunisia, where wood is exploited in the façades and doors (*Photo credit: Mohsen Aboulnaga*)

Fig. 4.24 NOI center in Bolzano, north of Italy (*Photo credit: Mohsen Aboulnaga*)

(a) (b)

Fig. 4.25 Timber used in the New Library of Alexandria (Bibliotheca Alexandrina), Egypt. (**a**) The main reading section of New Library of Alexandria (*Photo credit: Mohsen Aboulnaga*). (**b**) Private reading area of New Library of Alexandria (*Photo credit: Mohsen Aboulnaga*)

Fig. 4.26 Wooden crafted cladding in Al-Maghara Church, Old Cairo—Cairo, Egypt (*Photo credit: Mohsen Aboulnaga*)

Fig. 4.27 Wooded crafted hanging corridor in the main church in Abbassiya—Cairo, Egypt (*Photo credit: Mohsen Aboulnaga*)

of timber material usage in the construction industry; it also aims at encouraging the sustainable experience of timber and broadens its usability and functionality in the industry.

4.3 Methodology

The method of the study is based on a comparative analysis approach, through review and assessment of old and contemporary experiences of integrating timber as a sustainable building material and providing relative analysis with significant criteria to evaluate the impacts on the built environment.

Fig. 4.28 Wood-crafted door at the main church in Abbassiya—Cairo, Egypt (*Photo credit: Mohsen Aboulnaga*)

Fig. 4.29 Wooded shutters in Florence, Italy (*Photo credit: Mohsen Aboulnaga*)

Fig. 4.30 Old building's wooded roof in Florence *(Photo credit: Mohsen Aboulnaga)*

Fig. 4.31 Wooden shutters and trees to improve liveability and indoor comfort in the two piazzas—center of Florence, Italy *(Photos' credit: Mohsen Aboulnaga)*

4.4 Benefits of Wood as a Sustainable Material

Through literature, the acoustical role of timber material in buildings is highly valued and investigated. A very good exemplary models are the opera building in Prague, Czech Republic [39], and Vienna Opera House and Vienna's Centre, Austria (Figs. 4.32, 4.33, and 4.34). A review by Caniato et al. highlighted the features of lightweight wooden elements and analyzed their performance in different walls and floors' typologies for their impact on noise and low-frequency sound insulation [40]. They utilized both subjective and objective means of study to highlight the noise reduction capability of timber. Also, the structural robustness of timber in buildings is highly acknowledged, and even tall buildings could be built using timber as a main structural element. Different timber construction methods are analyzed for their related sturdiness and failure performance.

Wood has been also used extensively in government buildings before the invention of aluminum window frames such as that of the Museum of Natural History in Vienna, Austria, as illustrated in Fig. 4.35a, b. In Barcelona, the Mediterranean wooden windows' frames and shutters are incorporated in the façades of the stylish buildings as depicted in Fig. 4.36a, b.

In addition, timber covered with gold is widely used in religious temples in Asia. Figure 4.37 presents some examples in Bangkok, Thailand. Moreover, lumber is also utilized in the interiors of many buildings in Astana, Kazakhstan, as shown in Figs. 4.38 and 4.39.

Additionally, Huber et al. [9] used deterministic approach to evaluate the structural role of timber. They concluded that lightweight timber frame's strength could be increased with the addition of simple rim beams. It is widely acknowledged that timber post and beam construction has similarities to the steel frame construction in burliness yet flexible enough to resist earthquakes as well [41]. Benefits of wood extend to features connected to the material life cycle, where the material is highly recyclable and reusable [42].

Fig. 4.32 The opera building in Prague, Czech (*Photo credit: Mohsen Aboulnaga*)

Fig. 4.33 The opera building in Vienna, Austria (*Photo credit: Mohsen Aboulnaga*)

Fig. 4.34 Lumber cladding inside Vienna Centre, Austria (*Photo credit: Mohsen Aboulnaga*)

(a) (b)

Fig. 4.35 The Museum of Natural History in Vienna, Austria (*Photo credit: Mohsen Aboulnaga*). (**a**) Wooden frames of the windows of the building's façade (*Photo credit: Mohsen Aboulnaga*). (**b**) A close view of the wooden windows' frames (*Photo credit: Mohsen Aboulnaga*)

(a) (b)

Fig. 4.36 Mediterranean windows' shutters and frames in the city of Barcelona, Spain (*Photo credit: Mohsen Aboulnaga*). (**a**) Wooden frames of the windows and their shutters, Barcelona (*Photo credit: Mohsen Aboulnaga*). (**b**) Wooden window's shutters and frames in a building in the Ramblas Street (*Photo credit: Mohsen Aboulnaga*)

4.5 Timber as New and Sustainable Material for Skyscrapers

Timber tower construction is very critical since it is used as a main construction material rather than concrete and steel. This does not make the building only more sustainable, but also lighter and less expensive and requires less material in the footings as well as having less carbon footprint [43]. Another factor involved is fire safety concerns which have been resolved using mass timber, e.g., cross-laminated timber (CLT), which is different from the traditional timber material. The mass timber is fire resistant such that fire causes the exterior surface to burn into charcoal

(a) (b)

Fig. 4.37 The Temple of the Emerald Buddha (Wat Phra Kaew) in Bangkok, Thailand (*Photo credit: Mohsen Aboulnaga*). (**a**) Wooden columns plated with gold (*Photo credit: Mohsen Aboulnaga*). (**b**) Golden ornaments of timber roof (*Photo credit: Mohsen Aboulnaga*)

(a) (b)

Fig. 4.38 The lounge area at Nazarbayev University in Astana, Kazakhstan (*Photo credit: Mohsen Aboulnaga*). (**a**) Timber cladding in the interior (*Photo credit: Mohsen Aboulnaga*). (**b**) Timber used in cladding columns (*Photo credit: Mohsen Aboulnaga*)

Fig. 4.39 Lumber cladding in the interior of one of the restaurants inside a shopping mall in Astana, Kazakhstan (*Photo credit: Mohsen Aboulnaga*)

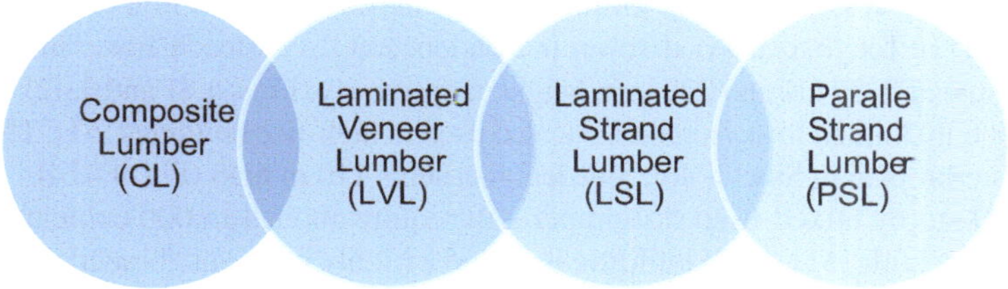

Fig. 4.40 Types of laminated timber and lumber composites for skyscrapers (*Source:* https://doi.org/10.1007/s00107-015-0999-5_)

and acts as a protective layer that creates a barrier for the inner timber layers and slows down the whole burning process (fire rating). This naturally happens in burning forests where the trees' trunks turn black while the inner layers are still alive. This property of mass timber is highly valuable to the construction industry [44]. Additionally, timber materials used in the construction of high structures are enhanced with added fire resistant and strengthening properties. These include manufactured wood products such as cross-laminated timber (CLT), and glued laminated timber (glulam), or lumber composites including composite lumber (CL), laminated veneer lumber (LVL), laminated strand lumber (LSL), and parallel strand lumber (PSL). These lumber composites are all considered sustainable wood products that are safe and tough for the skyscrapers' construction [44]. Figure 4.40 presents these types for skyscrapers.

With special emphasis on the high performance of CLT in skyscrapers' construction, the material is considered relatively strong, lightweight, and favored for being suitable for prefabrication, where it could be assembled in situ easily, hence resulting in a reduction in the total construction wastes, and thus is considered green. The CLT material involved in the construction of variety of building types such as multi-housing or commercial structures could be used for various building elements like walls and ceilings [45]. Another significant aspect of skyscraper's timber construction is the ability of timber to become highly stable to both earthquake movements and wind lateral forces. Advanced timber structures such as CLT are characterized with flexibility features that enable the material to swing in response to an earthquake, instead of being worn out by the force. The stability feature of timber in earthquakes is tested in Japan for high wooden-framed building tower that was able to survive an earthquake with 7.5 magnitude [46].

One of the well-known highest timber-built structures is a 55 m dormitory building over 18 storeys high established in 2016 in the United States. Another high-built timber structure is called "Treet", which is 53 m (14 storey) established in 2015 in Norway. While each is unique in design and construction, they share a significant common feature of the built super-high mass-timber structures [47]. Today, the capabilities of mass timber in the construction of skyscrapers have come to light. More proposals and plans to build timber and high-rise skyscrapers are being studied for construction.

The effectiveness and sustainability of timber, as a main building material, revolute the built industry, where more skyscrapers are to be built either fully or partially made of timber [48–50]. It is compelling to mention the highest proposed timber skyscrapers, W350 tower in Tokyo, Oakwood Tower in London, and River Beech Tower in Chicago, that measure 350, 304.8, and 228 m in height, respectively (Figs. 4.41 and 4.42). Timber use is not only limited to skyscrapers but also used in high-rise buildings [51]. The HSB 2023 Vasterbroplan in Stockholm, Sweden, measures 110 m high (Fig. 4.41d).

The 70-storey mixed-used skyscraper will require about 185,000 cubic meters of lumber to be built [51]. Such building is called a highly efficient "braced tube structure" which is 90% wood. Based on Sumitomo Forestry, the W350 skyscraper would impound 100,000 tons of CO_2 [52]. Moreover, mass timber, engineered timber products, and innovations associated with the construction and design of timber natural material create appealing opportunities for the use of timber in building

Fig. 4.41 Timber planned skyscrapers and high-rise buildings globally. (**a**) W350 tower in Tokyo, Japan (*Image source:* https://www.archdaily.com/889142/japan-plans-for-supertall-wooden-skyscraper-in-tokyo-by-2041). (**b**) Oakwood Tower in London, the United Kingdom (*Image source:* https://www.archilovers.com/projects/182005/oakwood-timber-tower.html). (**c**) River Beach Tower, Chicago, the United States (*Image source:* https://www.ft.com/content/0223bdec-801d-11e6-8e50-8ec15fb462f4). (**d**) HSB 2023 Vasterbroplan in Stockholm, Sweden (*Image source:* https://www.skyscrapercenter.com/building/hsb-2023-vasterbroplan/16172)

Fig. 4.42 Features of the timber skyscraper W350 in Tokyo, Japan (*Source:* https://resources.realestate.co.jp/living/70-story-wooden-skyscraper-planned-for-central-tokyo/). (**a**) Bird's-eye view of Tokyo's timber skyscraper W350, Tokyo, Japan (*Photo credit: Sumitomo Forestry*). (**b**) Front view of the laminated lumber/timber-structured skyscraper (*Photo credit: Sumitomo Forestry*). (**c**) Details of the timber elements of the main façade of the skyscraper (*Photo credit: Sumitomo Forestry*)

high-rise structures. These material advancements are considered the future of sustainable timber skyscrapers in the world; besides, they endorse both environmental and economic stewardship. Nevertheless, in the use of timber as to "Not to ignore" the capability of timber to absorb CO_2 emissions, timber skyscrapers are forecasted to reshape the construction industry by trapping harmful emissions.

4.6 Sustainable Contemporary Global Examples

Current global examples are presented and analyzed to evaluate the sustainability of timber usage in each case study. This includes three examples of building built with timber and lumbers. Figure 4.43 summarizes these three examples in Switzerland, Spain, and Indonesia.

Fig. 4.43 Global examples built with timber in Switzerland, Spain, and Indonesia (*Images credit:* https://m.dw.com/en/ecopia-intelligent-building-sustainable-living/a-16466122-0)

Sustainable Wood in the Alpine Shelter (Monte Rosa Hütte), Alps

On the top of the Alps mountains of Switzerland, an alpine shelter is built to set an example of a climate-friendly structure known as the Monte Rosa Hut/Monte Rosa Hütte, which is made of wood and cladded with aluminum to protect it from frost. Built in 2010, the shelter is designed for mountain hikers; thus, it is solely reachable on foot. The design concept is derived by "Swiss architect Andrea de Pras in cooperation with ETH Zurich the Swiss Federal Institute of Technology" to create a lodge suitable to accommodate mountain hikers in the cold climatic conditions of the region [53]. The biggest challenge was the ability to construct on an altitude of 3000 m. Also, the building couldn't be laid on a simple flat topographic land, but on the smallest footprint land area (Fig. 4.44). Thus, creating a spacious interior with the smallest footprint was another challenge. The project was then created on an octagonal plan with oblique exterior surface that opens upward; it was considered a very unique mountain loft [53].

Usage of Wooden Materials/Challenges of Wood Usage

The Monte Rosa Hütte is composed of a huge wooden staircase structure that ascends on the building peripherals close to the windows in a spiral setting that provides a panoramic view as shown in Fig. 4.45. The wooden staircase concept is mainly about creating a heat trap space where the sun penetrates directly into. The trapped heat is then distributed through the building spaces through the aerodynamics concept. The provided ventilation system treats the staircase well as a pipe that transports heat to other building spaces. Sun warmth is considered a natural heating system for building occupants in this extreme weather conditions. The sustainable

Fig. 4.44 Octagonal-shaped base chalet on top of mountain in the Alps, Switzerland (Image credit: https://www.youtube.com/watch?v=76KYLgz-BcQ)

heating concept does not only attract hikers but also curious visitors. Also, the interior is mostly composed of prefabricated wooden partitions for being a sustainable thermal and auditory insulating material [54].

Assessment of Sustainable Usage

The Monte Rosa Hütte is considered the most complex and expensive wooden structure in Switzerland. This is due to its location on top of the Alps and the fitted weather data center connected with ETH Zurich, the Swiss Federal Institute of Technology, to monitor its energy performance and stored energy generated from PV panels and solar water heating systems. The building altitude of this iconic structure is very challenging when it comes to providing building elements to the site. A helicopter had to fly over with all the building elements, which is considered a non-sustainable challenge and forms the only negative impact of the built structure to the environment. All structural elements were prefabricated in the valley and brought to assemble on site as shown in Fig. 4.45a, b. Another challenge is the permafrost nature of the climate; thus, stainless steel foundations are utilized in the construction of Monte Rosa Hütte to protect it in the freezing weather.

A strong sustainable aluminum shell wraps the building structure protecting it from extreme weather, related moisture problems, frost, and melting ice. The shell cover is designed to keep the interior heat and avoid the exterior humidity condensation on the façade surface. All of those contribute to the increase in the construction cost amount to a total average of 4.5 million euro [55].

Fig. 4.45 *Building is surrounded with timber staircase that traps heat, Switzerland.* (**a**) A wooden staircase acting as a thermal solar chimney to heat the interior (Source and images' credit: https://www.youtube.com/watch?v=76KYLgz-BcQ). (**b**) Wooden panels with insulation material dividing the octagonal plan. (**c**) The wooden skeleton appearing during the construction of the building from inside the building

Mega Wood Waffles in the Metropol Parasol in Seville, Spain

The largest wooden waffle structure named "Metropol Parasol" is located at the Plaza de la Encarnación in Seville, Spain, to be a cultural and commercial hub in the city of Seville (Fig. 4.46). The project which is composed of a museum and a commercial mall in the underground floor of the wooden structure is designed by J. Mayer H. Architects to create a coverage structure of 150 m long by 70 m wide and an average of 26 m in height [56]. The project is constructed using wooden panels that interweave into a waffle structure with reinforced concrete bases. Such

Fig. 4.46 Sustainable and recycled timber used in the Metropol Parasol, Seville, Spain (*Image credit*: https://www.dezeen.com/2011/04/26/metropol-parasol-by-j-mayer-h/)

wooden panels are made of recycled wood. The wooden structure forms convenient canopies and walkways, but in addition, the Metropol Parasol includes a wide range of amenities for the public usage [57].

Usage of Wooden Materials

The complete iconic structure is composed of baffled wooden panels that allowed a wide coverage area with minimum posts. In this building, timber is considered the main structural element, where cut timber elements named "Kerto laminated veneer lumber (LVL) Q-panels" are laid and glued in 1.5 m by 1.5 m orthogonally and fixed perfectly with additional steel connectors. The Kerto panels are well known for their high strength and light weight, characterized with unique strength to weight ratio. More than 3000 Kerto wooden elements are assembled to cover over 2500 m^3 of total LVL with weather-protective layer [56]. Figure 4.46 shows the Kerto timber assembled elements.

Assessment of Sustainable Usage

The Metropol Parasol is not entirely made of wood, but steel and reinforced concrete were both used as construction materials, beside the baffled timber structures, especially in post construction, to withstand subject loads [57]. The construction costs were 50 million euro (2005–2007)—it is considered very expensive at that period. However, the structure is built with large spanned gaps to avoid disturbing the local roman ruins within the site. Six mushroom columns form the structure to properly support the wide span with embedded stairs and lifts. Polyurethane foam seal is applied to glue the mega structure into place, especially for its capability to resist hot summer in Spain [57]. Figures 4.47 and 4.48 illustrate the features of the Metropol Parasol in Seville.

Fig. 4.47 Metropol Parasol wooden elements and mushroom columns in Seville, Spain (*Source and images' credit:* https://www.publicspace.org/works/-/project/g315-meteropol-parasol*)*. (**a**) The entrance of the building with timber panels spread all over. (**b**) The timber mushroom columns with large span appearing. (**c**) One of the sides of the assembled LVL timber elements. (**d**) Bird's-eye view of the timber Kerto panels near the visitors' bridges. (**e**) One of walking bridges above the wooden Kerto LVL panels. (**f**) Close view of the mushroom column near the walking passage

Recycled Wood in the Allele Aviles Mula Watteau Resort in Bali, Indonesia

Bali resorts are well known as the most touristic destinations for holiday seekers. As a result, nowadays, Bali suffers from rising environmental-related problems such as high traffic emissions and high levels of wastes [58]. The US Environmental Protection Agency (EPA) highlighted the significance of resources conservation in

Fig. 4.48 Details of Kerto LVL Q-panels with weather-protective layers in the Metropol Parasol, Seville, Spain (*Image credit:* https://www.publicspace.org/works/-/project/g315-meteropol-parasol)

the region [59]. A resort located on the southern end of Bali, on Bukit Peninsula, named "allele Aviles mula Watteau" follows high-sustainability stewardship to the environment. The hotel is planned by "Whoa-ha" architectural firm with a climate-based approach with open master plan and outdoor walkways. The aim is reducing operational energy by creating comfortable spaces with natural ventilation and reducing embodied energy by integrating sustainable materials. These site and buildings managed to integrate natural materials such as local limestone for façade cladding, loosely piled volcanic pumice for roof insulation, and recycled wood for roof and wall panels as shown in Fig. 4.49.

Usage of Wooden Materials

The sustainable wood integrated in the resort is recycled tropical wood that has been previously used in local railways and local telephone posts and telephone cabanas [45]. Using local recycled wood in the resort complex provides another strong dimension in region attaining sustainability goals [59]. In the resort complex, various forms of wooden elements are integrated and reflected in both outdoor and indoor settings. However, the recycled wood panels are used for the façade cladding, open pergolas, and outdoor features as shown in Fig. 4.50.

Assessment of Sustainable Usage

The sustainable usage of various natural materials in the resort complex emphasized on the recycled wood which aims at solving a rising problem of depletion of Bali resources as illustrated in Fig. 4.50. Local resources are selected to ensure highly sustainable integration of resources. Natural limestone is to be locally provided from the resort quarry, while volcanic pumice is gathered from the nearby Java Island. The sustainable approach extends to cover local workers with 60% Balinese people and educate these workers about sustainable awareness. Sustainability of

Fig. 4.49 *Natural materials in the allele Aviles mula Watteau resort in Bali, Indonesia* (Image credit: https://www.youtube.com/watch?v=76KYLgz-BcQ)

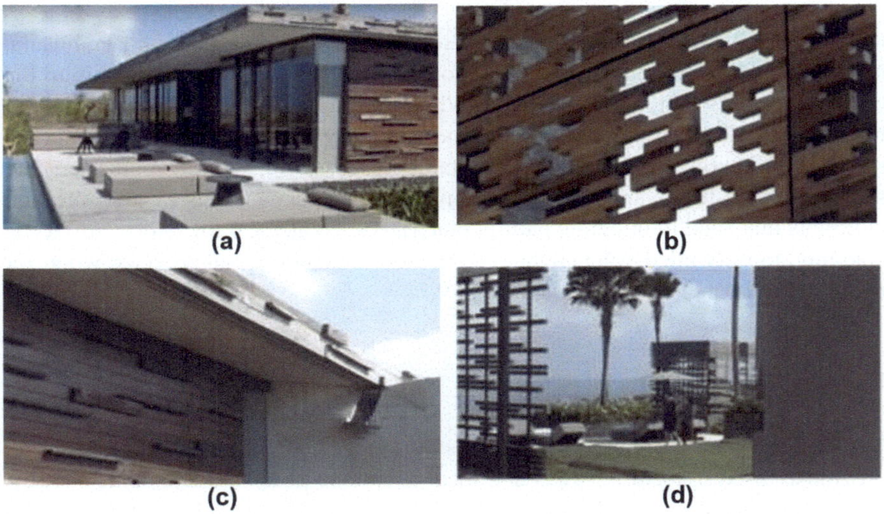

Fig. 4.50 *Usage of recycled wooden elements, the allele Aviles mula Watteau resort in Bali, Indonesia* (Image source: https://www.youtube.com/watch?v=76KYLgz-BcQ/). (**a**) Cladded railway wooden panels. (**b**) Wooden screen on the façades. (**c**) Wooden panels fixed on the wall. (**d**) The wooden screen panels

water resources is also attained through rainwater harvesting and 80% recycled water usage. Thus, the resort sets a highly sustainable example for both the conservation of resources and sustainable education. Also, the sustainable usage of materials in the resort complex allows it to function as an integrated part of the surrounding landscape [60]. To sum it up, the resort is considered as an effective component of a sustainable economy not just a separate complex.

4.7 Old Examples of Wooden Buildings in Egypt

This section of the study presents old examples of building cases with timber integration in construction in Egypt. This includes the Khedivial Opera House in Cairo and old wooden verandas and timber buildings in Port Said, Egypt.

Khedivial Opera House in Cairo

The old opera house is one of the oldest opera houses in Africa, in which Khedive Ismail ordered to build in 1861 to celebrate the Suez Canal opening by then. It was built by the architects Pietro Avoscani and Rossi to seat 850 persons [61], and the building was entirely made of wood as shown in Fig. 4.51. The opera, which was designed in both Baroque and Rococo styles that reflected the pure western-built architecture, was considered a huge artifact with the aid of Italian artisans to decorate the house interior and exterior to provide such a fascinating experience [62].

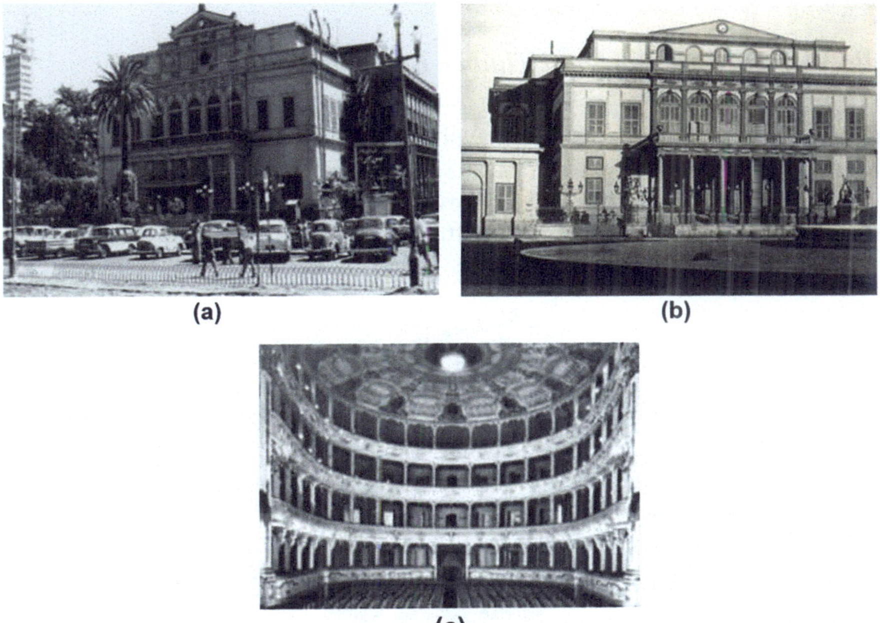

Fig. 4.51 External and internal views of the 1869 Khedivial Opera House in Cairo, Egypt. (**a**) The old opera house building in Ibrahim Pasha Square, Cairo (*Image credit:* https://www.cairoopera.org/history.php?lan=En). (**b**) Old royal opera house main façade, Cairo (*Image Credit:* https://en.wataninet.com/culture/heritage/the-splendid-opera-house-ismail-built-150-years-on/31062). *(**c**) An interior of the royal opera house, Old Cairo, Egypt* (Image credit: https://en.wataninet.com/culture/heritage/the-splendid-opera-house-ismail-built-150-years-on/31062)

Usage of Wooden Materials

The Khedivial Opera House was built with limestone foundation and mostly made of non-treated natural wood, which did not allow the structure to resist severe conditions. Carved wooden decorative panels are used all over the building structure to create the royal opera artistic effect as intended [63].

Assessment of Usage

The timber structure was not fire resistant, and the low specifications of the utilized wood participated in an enormous ruin of the material hub in a huge fire that started in October 1971. The wooden building was entirely burnt down in 10 h, and the Khedivial Opera couldn't be saved from the fire. Currently, the site building is replaced with a concrete multistorey car park, and site is named Opera Square (*Meidan el-Opera*). Nonetheless, Egyptians though knew that the burnt Opera is irreplaceable, and in 1988, a new opera house is built and named Cairo Opera House near Nile River in Cairo [64].

Old Wooden Verandas and Timber Buildings in Port Said

Most of old buildings, originally built with beautiful wooden verandas in the city of Port Said, Egypt, on the Mediterranean Sea, craft and depict an era of rich culture and architecture. Nearly all these built wooden façades' features are very old and may date back to a complete century [65] as shown in Fig. 4.52. Splendid and

Fig. 4.52 Virginia Arvanitopoulos Building, built in 1891 at the corner of al-Geish and Ahmed Shawki streets in the city of Port Said, Egypt. (**a**) The wooden building from two sides (*Image credit:* https://tayaramuse.com/a-walkthrough-of-what-to-see-in-port-said-egypt). (**b**) Wooden façade of the building with verandas (*Image credit:* https://rawi-magazine.com/articles/portsaidverandas)

Fig. 4.53 Splendid and spectacular wooden staircases embellish many of the buildings like Virginia Arvanitopoulos Building on al-Geish Street in Port Said (*Image Credit:* https://rawi-magazine.com/articles/portsaidverandas)

Fig. 4.54 Petrovich building, designed by French architects in one of the streets in the city of Port Said (*Image credit:* https://rawi-magazine.com/articles/portsaidverandas)

magnificent staircases decorate the interiors of many of these wooden buildings in Port Said as illustrated in Fig. 4.53. Also, Fig. 4.54 presents the Petrovich building, which is built from wood by a France architect.

Moreover, some examples of wooden buildings and verandas also built in the city of Port Fouad on the other side of Suez Canal are presented in Fig. 4.55.

Usage of Wooden Materials

These buildings are characterized with low height (3–4 levels) with additional protruded wooden balconies or verandas. The city of Port Said became famous with these unique buildings' features. However, originally, these verandas were built to protect the building façade from direct sunlight as adapted from similar Spanish, Portuguese, and French buildings in early nineteenth century. These neoclassic-styled façades in the city of Port Said, northeast of Egypt, are modified with wood (mashrabiya) to persuade the local cultural and environmental needs [66].

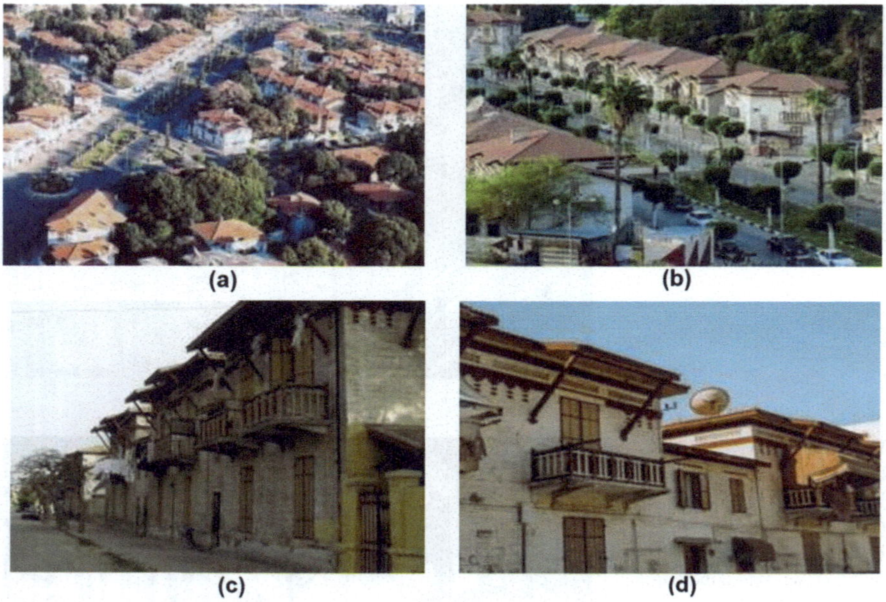

Fig. 4.55 Wooden villas and terrace houses in the seaside city of Port Said and city of Port Fouad, north of Egypt. (**a**) Bird's-eye view of the beautiful city of Port Fouad (*Source:* https://www.travelstart.com.en/blog/10-unerrated-egyptian-cities-worth-visiting). (**b**) Some of wooden villas for Suez Canal staff in Port Fouad (*Source:* https://www.printerset.com). (**c**) The wooden terrace houses with wood verandas and roofs, Port Said (*Image credit:* https://www.printerset.com). (**d**) Buildings' façade with wooden roofs and verandas in Port Said (*Image edit:* https://tayaramuse.com/a-walkthrough-of-what-to-see-in-port-said-egypt)

Assessment of Usage

The city encounters explicit decaying timber in the old timber-built verandas, and these are associated with the existence of the historical buildings that is vulnerable to demolition and threatened every day. Thus, possible loss of the city of Port Said architectural identity should be prevented with careful treatment of old timber in the building façades. Still, dominant building examples are surviving in the center of Port Said such as those built in the nineteenth century by Swiss architect Alberti, including the Slavic Building, the Coroni Building, and Virginia Arvanitopoulos Building as well as Petrovich building [67]. These verandas are considered Port Said cultural heritage; they are specified with variety of high-quality decorative façade features. The Protection of Cultural Heritage Association (TPCHA) in the city of Port Said successfully managed to register 505 historical buildings of the city for protection with a strong spirit to preserve these old neglected structures in the city [68].

4.8 Comparative Analysis of Examined Case Studies

A comparative analysis is conducted on all the previous assessed cases. The assessment criteria which include the building state, type, and timber role are analyzed and compared. It also summarizes the studied examples and emphasizes the role of timber in old and contemporary buildings, in order to highlight the significance of timber as a sustainable building material. The assessment encompasses four sets of status: (a) old cases, (b) new built cases, (c) new cases, and (d) planned cases. Table 4.1 lists the analysis findings.

The findings reveal that timber usage is mostly utilized in all buildings for the multirole characteristics of the material whether in buildings' skeleton, façades, and roofs as well as in the interior and furniture. Timber is also utilized for its excellence in many environmental issues such as acoustical and thermal as well as structural, artistry, and recycled performance in different buildings.

It is clear from Table 4.1 that the integration of timber in both old and contemporary buildings, as main structural building elements, is considered the most significant role across the analyzed cases (17 cases of the 26 assessed examples). Then, the thermal role of timber is integrated in building cases specifically located in tropical and Mediterranean countries that require high protection from heat (13 cases). The material is less frequently integrated in buildings for the acoustical role (five cases), the artistry role, and the recycled role exempted in building examples. However, the capabilities of timber as a sustainable material are not yet fully integrated or supported in current building examples and expected to exceed in performance in sustainable future buildings.

4.9 Conclusions

Timber is one of the oldest building construction materials in history; it has been cherished for its wide range of valuable characteristics and benefits in construction. Throughout history, colossal number of examples manifested the use of wood/timber in their skeleton, façade cladding, or verandas. Timber has been used extensively in Europe, North America, and Asia as well as in ancient Egypt. With everyday technologies and advancements, engineers discovered that there are no more sustainable building materials than those that exist naturally in the environment. From old to contemporary experiences, lessons are learned, and advancements are invented to highlight the sustainability of timber in the built industry. Further, the aim is a built environment that is entirely made of natural materials, effective, and highly sustainable.

Timber has proven to be not only an excellent insulation and building material but also a sustainable material. Recently, timber—in the form of laminated mass timber (lumber)—has been used to construct skyscrapers in many global large mega cities like Tokyo, London, Chicago, and Stockholm. Such new technology is

Table 4.1 Comparative analysis of timber material in examined case studies

Status	Name	Timber features		Design Role Interior – Exterior		
		Acoustics	Thermal barrier	Structural element	Carved elements	Recycled material
Old	Khedivial Opera House	□□	□□	□□	□□	
Old	Wooden verandas buildings in Port-Said		□□	□□	□□	
New Built	Metropol Parasol		□□	□□		
New Built	Allele Aviles Mula Watteau resort					□□
New Built	The Alpine shelter known as the Monte Rosa Hutte	□□	□□	□□		
Planned	W350 Tower in Tokyo, Japan			□□		
Planned	Oakwood-timber-tower			□□		
Planned	River Beach Tower			□□		
Old	Notre Dame de Paris, France			□□		
Old	Market Square buildings in Brussels, Belgium			□□		
Old	Royal garden in Vienna, Austria			□□		
Old	The Dome of the Duomo in Florence, Italy			□□		
Old	The Royal Dusit Palace (summer residence) in the outskirts of Bangkok, Thailand			□□		
Old	The old royal building in Seoul, South Korea			□□		
Old	The Jiming temple in Nanjing China			□□		
New Built	Screens in Nazarbayev University in Astana, Kazakhstan		□□			
Old	windows' shutters in Rome		□□			
Old	Pordenone, and in Bolzano, South Tyrol, Italy		□□			
Old	Royal building wooden shutters, Schönbrunn Palac in Vienna, Austria		□□			
New	building wooden façade in Spain		□□			
Old	Window shutters in Abdeen Palace, Cairo		□□		□□	
Old	buildings' façades in Casablanca, Morocco		□□			
Old	building in Sidi-bu Saeed, Tunisia		□□	□□	□□	
New	The New Library of Alexandria, Egypt	□□	□□			
Old	Hanging church (Al-Maghara) church in Cairo	□□		□□		
Old	Opera Building in Prague Czech Republic	□□		□□		

considered a paradigm shift in the sustainability of buildings and cities to attain sustainable development and SDGs.

References

1. Abo-elazm, F. M., & Ali, S. M. (2017). The concept of "Local Smart Architecture": An approach to appropriate local sustainable buildings. *International Journal of Cultural Heritage, 2*, 1–12. Retrieved June 15, 2020, from https://www.iaras.org/iaras/journals/caijch/the-concept-of-local-smart architecture-an-approach-to-appropriate-local-sustainable-buildings/.
2. Orsini, F., & Marrone, P. (2019). Approaches for a low-carbon production of building materials: A review. *Journal of Cleaner Production, 241*, 118380. https://doi.org/10.1016/j.jclepro.2019.118380/. Retrieved June 15, 2020.
3. Layla, N., Liebe, E., Bodaghi, E., Prajapati, T., Hisham, M. T. M., Abdelaziz, M. B., & Moyano, G. R. L. (2019). Sustainable construction. *Sustainable cities II*, 13. Retrieved June 17, 2020, from https://www.theseus.fi/bitstream/handle/10024/260737/2019_TAITO_28_Polloc k_Sustainable%20cities%20II.pdf?sequence=5&isAllowed=y#page=13/.
4. Charkovska, N., Halushchak, M., Bun, R., Nahorski, Z., Oda, T., Jonas, M., & Topylko, P. (2019). A high-definition spatially explicit modelling approach for national greenhouse gas emissions from industrial processes: Reducing the errors and uncertainties in global emission modelling. *Mitigation and Adaptation Strategies for Global Change, 24*(6), 907–939. https://doi.org/10.1007/s11027-018-9836-6/. Retrieved June 20, 2020.
5. Li, Y. L., Han, M. Y., Liu, S. Y., & Chen, G. Q. (2019). Energy consumption and greenhouse gas emissions by buildings: A multi-scale perspective. *Building and Environment, 151*, 240–250. https://doi.org/10.1016/j.buildenv.2018.11.003. Retrieved June 30, 2020.
6. Cavalli, A., Cibecchini, D., Togni, M., & Sousa, H. S. (2016). A review on the mechanical properties of aged wood and salvaged timber. *Construction and Building Materials, 114*, 681–687. https://doi.org/10.1016/j.conbuildmat.2016.04.001/. Retrieved September 7, 2020.
7. Forest Products Laboratory (US). (1987). *Wood handbook: Wood as an engineering material*. The Laboratory. Retrieved September 7, 2020, from https://books.google.com.eg/books?id=gwcUAAAAYAAJ&ots=GHcTEnNGhC&dq=However%2C%20in%20North%20America%20%E2%80%93%20Canada%20and%20the%20United%20States%2C%20the%20word%20%E2%80%98timber%E2%80%99%20refers%20to%20wood%20from%20felled%20trees.%20Lumber%20utters%20products%20from%20wood%20or%20timber%20&lr&pg=PP3#v=onepage&q&f=false/.
8. Braulio-Gonzalo, M., Gallego Navarro, T., & Lecha Sangüesa, A. (2014). Analysis of timber as sustainable material for construction. In *International Conference on Innovative Technologies, INTECH 2014*, (Leiria, Portugal), 10-13.09.2014 (pp. 89–92), University of Rijeka. Retrieved September 7, 2020, from http://hdl.handle.net/10234/188060/.
9. Huber, J. A., Ekevad, M., Girhammar, U. A., & Berg, S. (2019). Structural robustness and timber buildings—A review. *Wood Material Science & Engineering, 14*(2), 107–128. https://doi.org/10.1080/17480272.2018.1446052/. Retrieved September 7, 2020.
10. Vannucci, P. (2020). *Structural study of the Notre-Dame ancient charpente*. Retrieved July 5, 2020, from https://arxiv.org/abs/2005.12584/.
11. Weitz, A., Charruadas, P., Fraiture, P., & Hoffsummer, P. (2016). Roof frames in the built heritage of Brussels (Belgium): The contribution of dendrochronology and written sources. Tree Rings in Archaeology, Climatology and Ecology, 14, Retrieved July 5, 2020, from Proceedings of the DENDROSYMPOSIUM 2015 : May 20th - 23rd, 2015 in Sevilla, Spain (uliege.be). https://orbi.uliege.be/bitstream/2268/197389/1/TRACEp14-22.pdf/.
12. Vandenabeele, L., Bertels, I., & Wouters, I. (2017). The timber roof of the gothic revival broodhuis (Brussels, 1873-1895): Design, construction and collaboration. In *Building Histories*

(pp. 213–224). : Retrieved July 5, 2020 https://books.google.com.eg/books?id=93kkDwAAQBAJ&lpg=PA213&ots=4S71WtUaUA&dq=The%20Timber%20Roof%20of%20the%20Gothic%20Revival%20Broodhuis&lr&pg=PA213#v=onepage&q=The%20Timber%20Roof%20of%20the%20Gothic%20Revival%20Broodhuis&f=false/.

13. Betteridge, C. Whispering Pines–an HP oser house with a Paul Sorensen garden at Blackheath. In *Blue Mountains* (pp. 21–33). Retrieved July 5, 2020, from http://bluemountainsheritage.com.au/wpcontent/uploads/2018/11/BMHJ17040617sm.pdf#page=27/.
14. Kozak-Holland, M., & Procter, C. (2014). Florence Duomo project (1420–1436): Learning best project management practice from history. *International Journal of Project Management, 32*(2), 242–255. https://doi.org/10.1016/j.ijproman.2013.05.003/. Retrieved July 5, 2020.
15. Askew, M. (1993). *The Banglamphu District: A portrait of change in inner Bangkok*. Bangkok : Thailand Development Research Institute Foundation. Retrieved July 5, 2020, from http://tdri.or.th/wp-content/uploads/2013/05/Y93L.pdf/.
16. Choi, J. D. (2010). The palace, the city and the past: Controversies surrounding the rebuilding of the Gyeongbok Palace in Seoul, 1990–2010. *Planning Perspectives, 25*(2), 193–213. https://doi.org/10.1080/02665431003613014/. Retrieved July 20, 2020.
17. Liu, Y., Cai, J., & Zhang, J. (2019). Research on characteristics of timber framing techniques of representative constructions by Xiangshan Bang in Nanjing. *E&ES, 233*(2), 022025. Retrieved July 20, 2020, from https://iopscience.iop.org/article/10.1088/1755-1315/233/2/022025/meta.
18. Arkar, C., Domjan, S., & Medved, S. (2018). Lightweight composite timber façade wall with improved thermal response. *Sustainable Cities and Society, 38*, 325–332. https://doi.org/10.1016/j.scs.2018.01.011/. Retrieved July 20, 2020.
19. Nu.edu.kz. (2021). Nazarbayev University. [online] Available at: [Accessed 12 April 2021]. https://www.nu.edu.kz/.
20. King, A. C., & Potter, T. W. (1990). A new domestic building-façade from Roman Britain. *Journal of Roman Archaeology, 3*, 195–204. https://doi.org/10.1017/S1047759400010965/. Retrieved July 20, 2020.
21. Lucchi, D. E. (2020) .*Learning from the past: The recovery and the optimization of the original energy be-haviour of "Portici" Houses in Bolzano*. Retrieved July 20, 2020, from http://www.eurac.eu/en/research/technologies/renewableenergy/publications/Documents/EURAC_RenEne_AICARR2014_DExner-ELucchi.pdf/.
22. Cuerda, E., Pérez, M., & Neila, J. (2014). Façade typologies as a tool for selecting refurbishment measures for the Spanish residential building stock. *Energy and Buildings, 76*, 119–129. https://doi.org/10.1016/j.enbuild.2014.02.054/. Retrieved July 20, 2020.
23. Krist, G., & Iby, E. (Eds.). (2015). Investigation and conservation of East Asian Cabinets in imperial residences (1700-1900): Lacquerware & Porcelain. In *Conference 2013 Postprints* (Vol. 11). Böhlau Verlag Wien. Retrieved July 20, 2020, from https://books.google.com.eg/books?id=gqulDAAAQBAJ&lpg=PA3&ots=bs6c8CGPCf&lr&pg=PA3#v=onepage&q&f=false/ .
24. Abdelmegeed, M. M. M. (2020). Documentation of construction systems, type of damages and modification processes in façades of unregistered heritage-buildings in Khedival Cairo, Egypt. *HBRC Journal, 16*(1), 77–112. https://doi.org/10.1080/16874048.2020.1779474/. Retrieved July 20, 2020.
25. Morton, P. (1998). A study in hybridity: Madagascar and Morocco at the 1931 colonial exposition. *Journal of Architectural Education, 52*(2), 76–86. https://doi.org/10.1111/j.1531-314X.1998.tb00259.x/. Retrieved July 20, 2020.
26. Khaldi, L. (2015). *Vernacular aesthetics in self-built housing in Tunis and Cairo. Cities, communities & territories/Cidades, Comunidades e Territorios, 31*. Retrieved July 20, 2020, from http://hdl.handle.net/10071/10744/.
27. Asdrubali, F., Ferracuti, B., Lombardi, L., Guattari, C., Evangelisti, L., & Grazieschi, G. (2017). A review of structural, thermo-physical, acoustical, and environmental properties of wooden materials for building applications. *Building and Environment, 114*, 307–332. https://doi.org/10.1016/j.buildenv.2016.12.033/. Retrieved July 20, 2020.

28. Koca, G. (2019). Evaluation of wooden structures. In F. Bianconi & M. Filippucci (Eds.), *Digital wood design* (Lecture notes in civil engineering) (Vol. 24). Cham: Springer. https://doi.org/10.1007/978-3-030-03676-8_30/. Retrieved July 20, 2020.
29. Shafer, A. (2015). Sacred geometries: The dynamics of 'Islamic' ornament in Jewish and Coptic Old Cairo. In *Sacred precincts* (pp. 158–177). Leiden: Brill. https://doi.org/10.1163/9789004280229_011/. Retrieved July 20, 2020.
30. Hidaka, T., & Beranek, L. L. (2000). Objective and subjective evaluations of twenty-three opera houses in Europe, Japan, and the Americas. *Journal of the Acoustical Society of America, 107*(1), 368–383. https://doi.org/10.1121/1.428309/. Retrieved July 20, 2020.
31. Fidler, J., Maxwell, I., & Ridout, B. (2000). *Timber decay in buildings*. London Taylor & Francis. https://doi.org/10.4324/9781315024783/. Retrieved July 20, 2020.
32. Gonzalez, J. M., & Morrell, J. J. (2012). Effects of environmental factors on decay rates of selected white-and brown-rot fungi. *Wood and Fiber Science, 44*(4), 343–356. Retrieved July 20, 2020, from https://core.ac.uk/download/pdf/10195248.pdf/.
33. Östman, B., Brandon, D., & Frantzich, H. (2017). Fire safety engineering in timber buildings. *Fire Safety Journal, 91*, 11–20. https://doi.org/10.1016/j.firesaf.2017.05.002/. Retrieved July 20, 2020.
34. Sharma, N. K., Verma, C. S., Chariar, V. M., & Prasad, R. (2015). Eco-friendly flame-retardant treatments for cellulosic green building materials. *Indoor and Built Environment, 24*(3), 422–432. https://doi.org/10.1177/1420326X13516655/. Retrieved July 20, 2020.
35. Hart, J., & Pomponi, F. (2020). More timber in construction: Unanswered questions and future challenges. *Sustainability, 12*, 3473. https://doi.org/10.3390/su12083473/. Retrieved July 20, 2020.
36. William McDonough W., & Braungart M. (2010). *Cradle to cradle: Remaking the way we make things* (p. 193). North Point Press. Retrieved September 4, 2020, from https://books.google.com.eg/books?id=KFX5RprPGQ0C&lpg=PP1&ots=irLqy2 ueDi&lr&pg=PP1#v=onepage&q&f=false/.
37. Reid, H., Huq, S., Inkinen, A., MacGregor, J., Macqueen, D., & Mayers, J. et al. (2004). Using wood products to mitigate climate change: a review of evidence and key issues for sustainable development (pp. 13–20). London & Edinburgh, The United Kingdom: The International Institute for Environment and Development (IIED), and the Edinburgh Centre for Carbon Management (ECCM). Retrieved from https://pubs.iied.org/sites/default/files/pdfs/migrate/10001IIED.pdf?.
38. Rametsteiner, E., & Simula, M. (2003). Forest certification—An instrument to promote sustainable forest management? *Journal of Environmental Management, 67*(1), 87–98. https://doi.org/10.1016/S0301-4797(02)00191-3A/. Retrieved September 4, 2020.
39. The State Opera Prague. (2021). Retrieved 12 April 2021, from https://www.pragueresidences.com/praguelocally/classical-music-theater/the-state-opera-prague/.
40. Caniato, M., Bettarello, F., Ferluga, A., Marsich, L., Schmid, C., & Fausti, P. (2017). Acoustic of lightweight timber buildings: A review. *Renewable and Sustainable Energy Reviews, 80*, 585–596. https://doi.org/10.1016/j.rser.2017.05.110/. Retrieved July 8, 2020.
41. Gurtner, Y. K. (2007). Crisis in Bali: Lessons in tourism recovery. In E. Laws, B. Prideaux, & K. Chon (Eds.), *Management tourism crises* (pp. 81–97). Abingdon: CABI. Retrieved July 8, 2020, from https://books.google.com.eg/books?id=5LOG4Of9meYC&lpg=PA81&ots=EyNx1LW7ov&dq=Crisis%20in%20Bali%3A%20Lessons%20in%20tourism%20recovery.%20Crisis%20management%20in%20tourism%2C%2081-97.&lr&pg=PA81#v=onepage&q&f=false/.
42. Pei, S., van de Lindt, J. W., Barbosa, A. R., Berman, J. W., McDonnell, E., Daniel Dolan, J., & Wichman, S. (2019). Experimental seismic response of a resilient 2-story mass-timber building with post-tensioned rocking walls. *Journal of Structural Engineering, 145*(11), 04019120. https://doi.org/10.1061/(ASCE)ST.1943-541X.0002382/. Retrieved July 8, 2020.
43. Suter, F., Steubing, B., & Hellweg, S. (2017). Life cycle impacts and benefits of wood along the value chain: The case of Switzerland. *Journal of Industrial Ecology, 21*(4), 874–886. https://doi.org/10.1111/jiec.12486/. Retrieved July 8, 2020.

44. Brandner, R., Flatscher, G., Ringhofer, A., Schickhofer, G., & Thiel, A. (2016). Cross laminated timber (CLT): Overview and development. *European Journal of Wood and Wood Products, 74*(3), 331–351. https://doi.org/10.1007/s00107-015-0999-5/. Retrieved July 8, 2020.
45. Foster, R. M., Reynolds, T. P., & Ramage, M. H. (2016). Proposal for defining a tall timber building. *Journal of Structural Engineering, 142*(12), 02516001. https://doi.org/10.1061/(ASCE)ST.1943-541X.0001615/. Retrieved July 8, 2020.
46. Green, M., & Taggart, J. (2020). *Tall wood buildings: Design, construction and performance*. Basel: Birkhäuser. https://doi.org/10.1080/24751448.2018.1497379/. Retrieved September 4, 2020.
47. Nguyen, T. T., Dao, T. N., Aaleti, S., van de Lindt, J. W., & Fridley, K. J. (2018). Seismic assessment of a three-story wood building with an integrated CLT-lightframe system using RTHS. *Engineering Structures, 167*, 695–704. https://doi.org/10.1016/j.engstruct.2018.01.025/. Retrieved September 4, 2020.
48. Buffi, A., & Angelini, G. M. (2019). Adaptive timber towers. An evolutionary prototype for the 21st century skyscraper. In *Digital wood design* (pp. 971–987). Cham: Springer. https://doi.org/10.1007/978-3-030-03676-8_39/. Retrieved September 2, 2020.
49. Foster, R., & Ramage, M. (2017). Briefing: Super tall timber–oakwood tower. *Proceedings of the Institution of Civil Engineers-Construction Materials, 170*(3), 118–122. https://doi.org/10.1680/jcoma.16.00034/. Retrieved September 2, 2020.
50. Sanner, J., Fernandez, A., Foster, R., Ramage, M., Snapp, T., & Weihing, D. (2017). River beech tower: A tall timber experiment. *CTBUH Journal*, (2), 40–46. Retrieved September 1, 2020, from https://api.semanticscholar.org/CorpusID:53073913/.
51. Sabogal, M., & Amram, A. (2017). *Tall mass-timber building* (Doctoral dissertation, Virginia Tech.). Retrieved September 9, 2020, from http://hdl.handle.net/10919/78297/.
52. Carvalho, L. F., Carvalho Jorge, L. F., & Jerónimo, R. (2020). Plug-and-play multistory mass timber buildings: Achievements and potentials. *Journal of Architectural Engineering, 26*(2), 04020011. https://doi.org/10.1061/(ASCE)AE.1943-5568.0000394/. Retrieved September 1, 2020.
53. Goymann, M., Wittenwiler, M., & Hellweg, S. (2008). Environmental decision support for the construction of a "green" mountain hut. *Environmental Science & Technology, 42*(11), 4060–4067. https://doi.org/10.1021/es702740f/. Retrieved September 1, 2020.
54. Dini, R., & Girodo, S. (2018). Shelters in the night. The role of architecture in the process of understanding high-altitude areas. *Journal of Alpine Research| Revue de géographie alpine, 106*(106-1). https://doi.org/10.4000/rga.3919/. Retrieved September 1, 2020.
55. Knoop, A. (2016). Integrating structural elements for innovative design. *Architect, 105*(12), 66–69. Retrieved September 1, 2020, from https://www.hanleywooduniversity.com/media/course/4562/story_content/exter nal_files/HWU_AR_Vectorworks_Dec_ADV16-DIGITAL.pdf/.
56. Haduch, B., & Marjański, Ł. (2017). Metropol parasol w Sewilli. *Kwartalnik Naukowy Uczelni Vistula, 4*(54), 105–113. Retrieved September 4, 2020, from https://www.infona.pl//resource/bwmeta1.element.desklight-419995f6-dc8d-4d40-805b-b62ba567ebc9/.
57. Koppitz, J. P., Quinn, G., Schmid, V., & Thurik, A. (2011). Metropol parasol—Digital timber design. In C. Gengnagel, A. Kilian, N. Palz, & F. Scheurer (Eds.), *Computational design modelling*. Berlin: Springer. https://doi.org/10.1007/978-3-642-23435-4_28/. Retrieved August 20, 2020.
58. Mason, P. (2015). *Tourism impacts, planning and management*. London: Routledge. https://doi.org/10.4324/9781315781068/. Retrieved August 20, 2020.
59. Cole, S. (2012). A political ecology of water equity and tourism: A case study from Bali. *Annals of Tourism Research, 39*(2), 1221–1241. https://doi.org/10.1016/j.annals.2012.01.003/. Retrieved August 20, 2020.
60. Hakim, L., Kim, J. E., & Hong, S. K. (2009). Cultural landscape and ecotourism in Bali Island, Indonesia. *Journal of Ecology and Environment, 32*(1), 1–8. https://doi.org/10.5141/JEFB.2009.32.1.001/. Retrieved August 20, 2020.

61. Barbour, N. (1935). The Arabic Theatre in Egypt. *Bulletin of the School of Oriental Studies, University of London, 8*(1), 173–187. Retrieved August 20, 2020, from https://www.jstor.org/stable/608109.
62. Mestyan, A. (2013). Power and music in Cairo: Azbakiyya. *Urban History, 40*(4), 681–705. Retrieved August 20, 2020, from https://www.jstor.org/stable/26398251/.
63. Mestyan, A. (2017). *Arab patriotism: The ideology and culture of power in late Ottoman Egypt*. Princeton: Princeton University Press. Retrieved August 20, 2020, from https://books.google.com.eg/books?id=aOtuDQAAQBAJ&lpg=PP1&ots=H9Ho4sztcM&dq=).%20Arab%20patriotism%3A%20the%20ideology%20and%20culture%20of%20power%20in%20late%20Ottoman%20Egypt.%20Princeton%20University%20Press%20photos&lr&pg=PP1#v=onepage&q=).%20Arab%20patriotism:%20the%20ideology%20and%20culture%20of%20power%20in%20late%20Ottoman%20Egypt.%20Princeton%20University%20Press%20photos&f=false/.
64. AlSadaty, A., ElKerdany, D., Hamza, N., Imam, S., ElSerafi, T., & Abdallah, M. (2020). Socio-spatial regeneration challenges in Attaba historic market, Cairo–Egypt. *Journal of Humanities and Applied Social Sciences*. https://doi.org/10.1108/JHASS-11-2019-0078/. Retrievd: August 20, 2020.
65. Megahed, N. (2014). Heritage-based sustainability in Port Said: Classification of styles and future development. *International Journal of Architectural Research: ArchNet-IJAR, 8*(1), 94–107. Retrieved August 18, 2020, from http://citeseerx.ist.psu.edu/viewdoc/download?doi=10.1.1.831.7897&rep=rep1&type=pdf/.
66. Piaton, C. (2017). Port Said: Cosmopolitan urban rules and architecture (1858–1930). In H. Abouelfadl, D. ElKerdany, & C. Wessling (Eds.), *Revitalizing city districts* (The urban book series) (pp. 3–14). Cham: Springer. https://doi.org/10.1007/978-3-319-46289-9_1/. Retrieved August 18, 2020.
67. Carminati, L. (2020). *The many boundaries within late 19th century Port Said. Controversial heritage and divided memories from the nineteenth through the twentieth century: Multi-ethnic cities in the Mediterranean World*, Volume 2, 92. Retrieved September 9, 2020, from https://books.google.com.eg/books?hl=en&lr=&id=-K7uDwAAQBAJ&oi=fnd&pg=PT38&dq=66.%09Carminati,+L.+(2020).+The+Many+Boundaries+within+Late+19th+century+Port+Said.+Controversial+Heritage+and+Divided+Memories+from+the+Nineteenth+through+the+Twentieth+Century:+Multi-Ethnic+Cities+in+the+Mediterranean+World,+Volume+2,+92&ots=UmJaj-VB1N&sig=yEgAFiJaRdGQspSqdZIbhXONgMw&redir_esc=y#v=onepage&q&f=false/.
68. Fouad, S. S., & Messallam, O. (2018). Investigating the role of community in heritage conservation through the ladder of citizen participation approach: Case study, Port Said, Egypt. *International Journal of Architectural and Environmental Engineering, 12*(11), 1539–1545. https://doi.org/10.5281/zenodo.2021845/. Retrieved August 20, 2020.

Chapter 5
Green Building Issues Using Wood and Timber in Buildings in the Arabian Gulf Countries

Falah Al-Kubaisy

5.1 Introduction

The agricultural areas in the GCC are scarce. Numerous sustainable environment research papers conducted by regional experts, who come from various backgrounds and fields, have confirmed that more than 90% of the total area in the Gulf states is unsuitable for agriculture. Geographically, the area is covered by deserts and seas without any rivers, streams, or forests in the surroundings. One of the critical challenges that hinders the sustainable development in these countries is the lack of renewable water resources. Rainfall in the Arabian Peninsula is rare and infrequent [2]. Hundreds of oases, scattered in the region, possess date palm trees which produce trunks and leaves used as the main roof structure of dwellings in rural areas. Meanwhile, timber and wood imported from Africa and Asia are utilized for windows, doors, and roofs to construct buildings throughout the last centuries up until the introduction of steel and concrete.

Traditional buildings in the Gulf cities and villages were designed to maximize shading, reduce thermal gain of the sun radiation, regulate building temperature, and enhance air circulation. These effects are achieved through a clever combination of building materials, placement, and design. Natural materials such as limestone, brick, and mud—in some cases mixed with local desert plants—provide a construction material with the capacity to regulate building temperatures. The material itself is capable of absorbing moisture in humid conditions, which can later evaporate during hot and sunny days to provide a slight cooling effect. Additionally, the sandy texture and color of the buildings reduce both the absorption and emission of radiating heat [3].

Many traditional structures feature an internal courtyard, often containing trees and a water well. The courtyard is typically surrounded by rooms or walls on all

F. Al-Kubaisy (✉)
Muharraq, Kingdom of Bahrain

A. Sayigh (ed.), *The Importance of Wood and Timber in Sustainable Buildings*, Innovative Renewable Energy, https://doi.org/10.1007/978-3-030-71700-1_5

sides, maximizing the area in shadow throughout the day and creating a space for socializing in the evenings. When the sun bears down at midday, the courtyard works as a chimney for the hot air to rise and become replaced by cooler air from the surrounding rooms—this promotes air circulation and creates a cooling effect (Figs. 5.1 and 5.2).

Fig. 5.1 Using timber and wood in new architecture in the Arabian Gulf is to cool the spaces by air movement and shading when having lattice windows or ceilings. Author collection. References: Left image is a private collection of the writer for Waqif Traditional Market in Doha - Qatar 2010. Right image is from seier+seier/Flickr

Fig. 5.2 Roof and wooden screen are placed in order to shade and make a comfort zone of traditional houses in Muharraq, Bahrain. Reference: Tarreq Wali, a book (published in Arabic) during 1990

Glass is not a common material in traditional buildings. A typical room has two external wooden windows: one would be an exceedingly small window located high up in the wall and is kept open to allow air to circulate and let in natural light. The second is larger, closed by wooden shutters and with grooves to allow the flow of air inside the room while maintaining privacy. Rooms also have windows positioned toward the internal courtyard for improved cooling.

Finally, a *Mashrabiya* or *Shanasheel*—a projected window overlooking the street with carved wooden latticework, typically located on the first floor of a building—allowed for better air circulation and view; see Fig. 5.3. Some buildings also have a wind tower, which creates natural ventilation by circulating cool air. The narrow streets allow them to be covered in most cases by light material from date palm trees to avoid direct sunlight. This allowed for better air circulation between streets and courtyards of buildings, via the rooms. All these features assisted in keeping traditional buildings cool. But the question remains: how can we apply them in today's cities?

In addition, a growing trend to the strength and quality issues affects the choice of wood products used in construction; the recent emphasis on sustainable building has also begun to have an impact on these developments. Making the right choice can help firms which aim to gain green certifications for their projects. Al-Habaibeh stresses that "Wood is the world's most abundant renewable resource. It's recyclable, bio-degradable and sustainable over the long-term." Looking out for TEFC[1] or FSC's[2] worldwide-recognized sustainability certification will become the norm in future [5].

Wood and timber have been used in construction projects for centuries. From some of the earliest buildings, it has featured prominently as a construction tool, a structural material, and an aesthetic finish. But what role is the material playing in

Fig. 5.3 Using wood extensively for windows and timber for main doors in traditional areas of Baghdad. Screen wooden panels enhanced the exterior facades of the building and made a shade to comfort interior spaces by allowing penetration of wind in the upper floors, as well as increase the privacy [4]

---

[1] A Totally Enclosed, Fan-Cooled electric motor is a type of industrial electric motor with an enclosure that does not permit outside air to freely circulate through the interior of the motor.

[2] Forest Stewardship Council.

today's sector? What benefits can be gained by using wood for construction, and what are the issues to consider? What benefits is this bringing to the end users, and which issues should they be aware of? Today, in the GCC countries, wood is the appropriate material as it increases the energy efficiency. Moreover, wood and timber could be considered as a material in construction in the Gulf region; wood is primarily used for two purposes—a formwork material during the construction process and to provide the final finish to the interior decor.

5.2 Design for Climate

The wind tower and *badgirs* [6] utilize the "Venturi Principle," that is, when funnelled from wider to a smaller aperture, the wind speed increases. A wind tower has four faces of about 7–10 square meters of open area. The central diagonal walls funnel the wind whatever its direction. The negative pressure on the leeward side draws air up from the room. At night when there is no wind, a "stock effect" is created whereby the hot air inside rises the tower and draws in cooler air from outside [7].

The *badgirs* may be either parapets to roof terraces or wall panels on the street side of rooms usually at first floor but sometimes on the ground. There are two *farsh* panels. The upper panel is about 15 cm inside the lower—the gap being purely horizontal. The upper panel is supported on a small pole spanning between coral block piers. On roof terraces, there may be deep coping across the top. The depth of the upper panel is 40 cm and the aperture 10 or 15 cm. The effect of these devices is quite startling—a refreshing little breeze is created even on a still day, and it is brought down to sitting or sleeping level (see Fig. 5.4).

Wall materials (gypsum, coral, and *farsh*) have low thermal capacity and have good insulating properties to let little heat pass. Rooms for summer nighttime were used on the roofs where the thin *farsh* panels lost their daytime heat very quickly. Ground-level walls are usually thicker. This kept the occupants warm on winter nights. A reflection is helped by pole colors and smooth surfaces. Absorption of heat may be cut to 20%, the remainder being reflected. Low thermal capacity and high reflectivity ensured that the wall surface did not heat up very quickly; it remained cool to touch and did not expand or crock. The incised decorative panels may have voids behind them, and this would also reduce the moss in the void, trap air, and truly reduce thermal capacity [9].

Wooden windows never have glass except for the fanlights—and that is stained to reduce glare and thermal transmittance. The screens allow air movement but decrease heat and glare and preserve privacy. There are shutters to retain daytime heat in the room on winter nights and to prevent high winds or driven dust when the shamal (northern winds) blow hard (see Fig. 5.5). Regarding orientation, two considerations were observed. An extremely long thin building seems best because it has maximum exposure to wind and the least of resistance to cross ventilation. If oriented with its long axis east-west, for instance, facing north-south, this will minimize heat gain from the sun. The prevailing wind is northwesterly, and so, one

Fig. 5.4 A wind tower in Muharraq, Bahrain [8]. Reference: Felibrilu/Flickr, CC BY-NC

Fig. 5.5 Timber and wood could be considered as a material in construction in the GCC region that increase efficiency of energy. Modern architecture inspired by heritage elements that wood and timber surge the intimate scale of the street in middle of Dubai, the UAE. Reference: Image from Gulf News, February 01, 2011, 00:00, by Liam Nelson, Features Writer, Property (https://gulfnews.com/business/property/why-wood-makes-sense-1.750875) [10]

would expect to find the *badgir* walls facing northwest or within 45° of this orientation for maximum benefit. However, these rules and regulations were not always strictly followed because of the privacy issues and due to people becoming more aware of the value of correct orientation around 1890. Moreover, wind flow was enhanced by the zigzag street form and the staggering of first floor apartments. But one suspects this is more due to chance than to planning.

Courtyard proportions are generally regarded as a key factor in control of temperatures, since they trap cold night air which continues to cool adjacent rooms for part of the day. In Bahrain, courtyards can be very large, except where lack of land and subdivisions of inherited property have led to densification. From this, one may conclude that small courtyards were not adopted as a deliberate design feature. The reason may have to do with the low diurnal variation—in other words, the air at night was not sufficiently cold to achieve much daytime cooling. Breeze may have been a more effective cooling agent, in which case larger courtyards could be big advantages.

5.3 Timber and Wood Sustain and Comfort Internal Spaces of Buildings

Increase Energy Efficient

Scanning the mega city's skyline like Dubai, Abu Dhabi, in the UAE or Doha in Qatar reveals a pastiche of ambitious but stylistically divergent architecture, a place where architects write their signature designs in a bold hand. This mélange gives the city an unmistakably modern yet varied aesthetics. Unifying this wide assortment of relatively new towers is a dependence on steel and glass that can be almost overwhelming in the summer months when the hot sun is reflected off these impressive facades.

Where steel and glass trail off, the eye is greeted by concrete or other masonry-type construction. Indoors, the average apartment or villa is a tiled vault that sends your voice echoing from room to room until the installation of furniture finally dampens the sound. There is a notable absence of a material that is ubiquitous in construction outside this region, and that material is wood. This being the case, a central concern of the GCC governments are sustainable and as such have become a component of design consideration throughout the country. However, energy efficiency might be an impediment to the use of wood products in construction. Architect Greg Mella, a principal at SmithGroup, who serves as a sustainability advisor to Masdar says, "Wood requires one-tenth of the energy required to make a steel beam of the same strength. That's because, to make the latter you must first collect scrap steel, transport it to the mill and melt it down at very high temperatures. With engineered wood products, you use fast growing, under-utilized and less expensive wood species. You don't have to cut down an old growth tree" [10]. The fact is where conventional lumber is required, in lieu of engineered wood products, there is a wealth of products available without the high environmental price tag. These are certified by the Forestry Stewardship Council (FSC), an independent, nongovernmental, and nonprofit organization established to promote the responsible management of the world's forests. The FSC certifiers monitor the supply chain from harvest to mill; in fact, use of FSC-certified products is one of the considerations in awarding LEED points to a project.

Application Benefits

In addition to its environmental properties, wood products have their own cosmetic and structural merits, suggests Thomas J. Westbrook, director of Middle East sales for Walsh Industries, a US-based firm which provides structural building materials for the international commercial construction market. "The real question is how and in what application wood is to be used in the UAE," says Walsh. But wood is aesthetically friendlier than steel or concrete, and it provides a diversity to the typical concrete and steel look and feel.

Structurally, wood can be used in many of the same applications as steel and concrete, particularly in the case of glulam timbers. Known formally as glued laminated timber, glulam is a type of structural timber product made from laminating smaller pieces of timber together. By combining smaller sections, the resulting product has improved tensile, compression, and bending strength allowing it to cover much longer spans and bear heavier loads than conventional timber [11] (see images in Fig. 5.6).

While wood products are less common in the region, other industries have seen their products successfully implemented in the GCC countries. Products can be used in roof systems to lighten the load typical to concrete and steel and in the form of radiant barrier oriented strand board (OSB) roof sheathing which, when placed under tile or shingle roof surfaces, reduces the amount of radiant heat entering the

Fig. 5.6 Using wood extensively for acoustic reasons in the auditorium of the National Theater in Bahrain to inspire a tradition that reflects over the contemporary architecture. Reference: Left image by Nicolas Buisson (http://www.nicolasbuisson.com); right image by Katrin Herden (https://www.pinterest.com/pin/573294227549945997/)

building by as much as 97%, thereby lowering heating, ventilating, and air conditioning (HVAC) load.

More specifically, glulam timbers can be seen in the executive car park of the Abu Dhabi Investment Authority's (ADIA) main office in addition to the Business Village—a project of the Mohammed Bin Rashid Establishment for Small and Medium Enterprises. The masterpiece provides both structurally and aesthetically superior shading [11]. The mass of concrete and glass buildings springs up around the modern area in the main cities of the Arabian region, as they currently provide overlay plywood panels which are imported to build the universal concrete forms that cover much of the developing urban landscape.

Sustainability

While timber and wood continue to be a superior building and construction material throughout the world, the application for wood products in the GCC countries goes well beyond the meagre applications currently being used. However, the largest impediment to the usage of wood in the GCC countries is the lack of understanding by the architectural, engineering, and project management sectors. Most of the people working in those sectors are very well informed about steel and concrete, but do not recognize wood.

The forming plywood offered by Wood Industries is engineered for an impressive reuse regiment, enabling multiple use of each panel up to 75 times or more. That level of reuse has a green attribute as well, since a longer useful life span per sheet means less plywood is used overall [11]. Contributing more directly to green building in the region, most firms currently provide "FSC products which qualify the user for sustainable building solutions, both under Estidama in Abu Dhabi, and to earn multiple LEED points throughout the UAE depending upon the product and application" [12].

Meanwhile, engineered wooden products have unique attributes that merit application-specific usage in any market. Also, one must look at the wooden brass lining in Dubai Creek to realize that wood construction already has a long-established niche in this region and is held in high regard both culturally and aesthetically. Whether or not aesthetics can make inroads in the Arabian Gulf region development industry remains to be seen (Fig. 5.7).

5.4 Wood Working

What benefits can be gained by using wood for construction, and what are the issues to consider?

Over the long term, the cost of wood is stable, and there is a continued demand for it. Wood is the world's most renewable source; it's recyclable, biodegradable,

Fig. 5.7 Multiuse of glulam timber as beams in friendly satiable building structure [11]. Reference: Glued Laminated Timber. Hasslacher https://www.hasslacher.com/en/from-wood-to-wonders

and sustainable. Timber was one material used during the construction of Ivory II at Business Bay, Dubai, first in the region followed by other. Wood has been used in construction projects for centuries. From some of the earliest buildings, it has featured prominently as a construction tool, structural material, and aesthetic finish. But what part is the material playing in today's sector, what benefits is this bringing to the contractors and end users, and of what issues should they be aware of?

Wood in Construction

In the Middle East today, wood is primarily used for two purposes—as a formwork material during the construction process and to provide the final finish to the interior decor. In other regions of the world, it is also used as a structural element, but this is generally an uncommon practice in the Gulf region. "Building codes and

Fig. 5.8 Doors and wooden ceiling that make a comfort zone of traditional reflect over the contemporary buildings are similarly used in the Arabian Gulf countries. Reference: Images from Mohammad Bin Al Zubair: Oman's Architectural Journey, Arabic book published in 2013 by BAZ Publishing, Sultanate of Oman

regulations differ from one country to another; the actual structures in the Middle East are almost always done with steel and concrete, but in areas such as North America and Europe, timber is used for the actual structure of the homes as opposed to concrete and steel," reports Madar Holding Purchasing Vice President Hesham El Abd. "The reasons [for this] are affordability, strength, design flexibility and beauty," he explains [13]. Some buildings in the Middle East do have wood implemented within the structure, including high-profile projects such as the Jumeirah Beach Resort in Dubai and Sharjah Sahara Mall; however, these remain largely as exceptions to the norm (Fig. 5.8).

The use of wood products remains mostly for aesthetics and to fit the general theme selected by the architects or designers. With forests not featuring greatly in the natural landscapes of the Gulf, most of the wood used for the construction industry must be imported, which raises issues of availability, cost, and transportation. However, the region's location does ease the potential problem of availability. The sources of wood vary according to the type, with supplies being offered from countries worldwide. Plywood is typically sourced from Indonesia, China, and Brazil; softwood timber from Chile, Austria, New Zealand, and Romania; and hardwoods from Africa, North America, and Indonesia to name but a few.

Market Forces

With the current global recession making costs even higher on the agenda than usual, can this material provide any respite to firms struggling to meet their project budgets? The need for wood is still there, even if there is no real construction boom.

Fig. 5.9 There are a lot of qualities in wood which have different grades. An unskilled person will not be able to identify these differences. The need for wood is still there, even if there is no real construction boom

The species used in the sector have seen the bottom and are now recovering. "Timber prices fluctuated widely for the last 2 years," reports Bhatia [14] (Fig. 5.9).

The continued demand for wood-based products is also ensuring that any short-term fluctuations remain relatively small. "Just like every other commodity in the world, wood prices have dipped over the past years or so, although not at the same rate as steel," reports El Abd. "The need for wood is still there, even if there is no real construction boom. People need to change their furniture, renovate their homes, put in hardwood flooring, etc. The wood species used specifically in the construction field have seen the bottom and are now recovering. In fact, certain species have recently seen price increases," he continued.

Procurement Issues

Before buying wood products, there are certain issues that contractors should consider—ranging from product quality and consistency to required treatments. "There are a lot of qualities in wood, which comes in different grades," explains Madhusudan Rao, head of sales, Danube Building Materials. Customers should educate themselves before selecting the species or specific grade needed for a project in order to ensure it will meet the build requirements. "One needs to understand the final use for the product in question and to take several factors into consideration such as the strength or density needed; climatic conditions; what grade is really needed and whether the material has to be [kiln dried]," explains El Abd. Whether wood needs to be treated prior to shipping to site or provided in its natural form is another matter that should be considered early in the procurement process, as this can affect delivery schedules. "Treatment for wood in general will depend on the application or end-use," explains El Abd. "Various treatment exists such as anti-stain treatment (Sinesto B) and kiln drying," he adds. Once kiln dried, timber will be stable and should not shrink or become susceptible to stains.

5.5 Conclusion

The strength and quality issues affecting the choice of timber and wood products used in construction and the recent emphasis on sustainable building are also beginning to have an impact on this sector. Making the right choice can help firms aiming to gain green certifications for their projects. Wood is the world's most abundant renewable resource. It's recyclable, biodegradable, and sustainable over the long term. It is not so strictly enforced here to request sustainable sources as of now, but soon, we expect that the government will levy such laws.

Wood is suitable for use in building construction with barely any restrictions [15]. This is new and requires creative rethinking of tried and tested practices in wood construction: classical categories can be replaced by mixed construction methods as necessary within a project, which yields completely new possibilities in designing wood structures. Manuals provide architects, engineers, and wood specialists with the essential expertise on the new systematics and construction methodology, from design, to prefabrication, to the implementation on site. It lays the grounds for mutual understanding among everyone involved in the project, to facilitate the necessary cooperation in the integral planning and construction process.

References

1. Mayo, J. (2015). *Solid wood: Case studies in mass timber architecture, technology and design*. Abingdon: Taylor & Francis.
2. Al-Rashed, M. F., & Sherif, M. M. (2000). Water resources in the GCC countries: An overview. *Water Resources Management, 14*, 59–75. Kluwer Academic Publishers. Printed in the Netherlands. 15 January 2000.
3. Al-Habaibeh, A. (2015). *Could traditional architecture offer relief from soaring temperatures in the Gulf?*
4. Author collection, Baghdad, Iraq.
5. Luke, A. (2009, April 18). What benefits can be gained by using wood for construction and what are the issues to consider? *Construction Week*.
6. Badgir: Is the Tower which tempts the Wind: A traditional Iranian architectural element to create natural ventilation in buildings. Wind-catchers come in various designs: Unidirectional, bidirectional, and multidirectional. The devices were used in ancient Iranian architecture. Wind-catchers remain present in Iran and can also be found in traditional Persian-influenced

architecture throughout the West Asia, including in the Arab states of the Gulf, West Iraq, Pakistan, and Afghanistan.
7. Yarwood, J., El-Masri, S., & Abd Alraouf, A. (2005). *Al Muharraq: Architectural heritage of Bahraini City*. Shaikh Ebrahim Bin Mohammed Al-Khalifa, miracle.
8. Felibrilu/Flickr, CC BY-NC.
9. Yarwood, J. p. 44 (same as Ref. [12]).
10. Nelson, L. (2011, February 1). *Gulf News*. By features writer, property. Retrieved from https://gulfnews.com/business/property/why-wood-makes-sense-1.750875.
11. Glued Laminated Timber. Hasslacher. https://www.hasslacher.com/en/from-wood-to-wonders. Glued laminated timber, also abbreviated glulam, is a type of structural engineered wood product constituted by layers of dimensional lumber bonded together with durable moisture-resistant structural adhesives. In North America, the material providing the laminations is termed laminating stock or lam-stock, see Wikipedia.
12. Chain of Custody, the FSC system allows businesses and consumers to identify, purchase and use wood, paper and other forest products made with materials from well-managed forests and/or recycled sources. The FSC label is found on wood and paper products as well as non-timber forest products such as latex. You can find the FSC logo on floors, decking, paper, printed matter, charcoal, kitchen utensils, venison and many more products!
13. Elabd, H. (2009, April 18). Chief Procurement & Operation Services Officer; King Abdullah Economic City (KAEC). Retrieved from https://www.arabianbusiness.com/wood-working-13537.html.
14. Bhatia, R. K. (2009, April). Arabian Business Journal, Industries.
15. Kaufmann, H., Kroetsch, S., & Winter, S. (2018). *Manual of multistorey timber construction*. Berlin: De Gruyter.

Further Reading

Michael Green by (Author) Jim Taggart "Tall wood buildings: Design, construction and performance". Birkhauser 2017.

Jones, S. (2018). *Mass timber: Design and research*. Oro Editions.

Kamariab, A., Corraoa, R., & Kirkegaardb, P. H. (2017). Sustainability focused decision-making in building renovation. *International Journal of Sustainable Built Environment, 6*(2), 330–350.

Kaufmann, H., Kroetsch, S., & Winter, S. (2018). *Manual of multistorey timber construction*. Berlin: De Gruyter.

Mayo, J. (2015). *Solid wood: Case studies in mass timber architecture, technology and design*. Abingdon: Taylor & Francis.

Menges, A., Schwinn, T., & Krieg, O. D. (2016). *Advancing wood architecture: A computational approach*. Abingdon: Taylor & Francis.

Mercader-Moyano, P. (Ed.). *The sustainable renovation of buildings and neighbourhoods*. Bentham eBooks: Indexed in: EBSC.

Sharifiand, A., & Murayama, A. (2013). A critical review of seven selected neighborhood sustainability assessment tools. *Environmental Impact Assessment Review, 38*, 73–87.

WBDG: Whole Building Design Guide: Historic Preservation Subcommittee. (2019). *Sustainable historic preservation*. Updated: 08-26-2019.

Zheng, H. W., Shen, G. Q. P., & Song, Y. (2017). Neighborhood sustainability in urban renewal: An assessment framework. *Environment and Planning B: Urban Analytics and City Science Journal, 44*, 903–992.

Chapter 6
Wood Building in Portugal

Luis Morgado, João Gomes Ferreira, and Manuel Correia Guedes

6.1 Against Wood

Masonry walls, firstly, and reinforced concrete structures, after, have dominated Portuguese construction practice, creating a routine and inhibiting wood construction from becoming a more natural and popular way of building. Context factors, such as the climate (temperate climate, with high temperatures in summer), the risk of biological attack by fungi and insects (species existing in Portugal), the scarcity of qualified wood (products graded for structural use), the lack of skilled and experienced technicians, the lack of incentives and support from the state, and the scarce national regulations on wood, seem to have been an obstacle to the process of "wood construction revival," which, in recent years, has animated many countries.

The Forest

With regard to the availability of raw materials, apparently, Portugal has the necessary conditions for a culture of wood construction to be consolidated. The national territory has a significant forest land cover: an area of 38.8% of the surface of the continental territory, according to the Statistics National Institute [1], which includes species that could supply wood for construction. However, until now, neither the domestic demand for structural timber has proven to be sufficient to boost industrial

L. Morgado · J. G. Ferreira · M. C. Guedes (✉)
Department of Civil Engineering and Architecture, Instituto Superior Técnico—University of Lisbon, Lisbon, Portugal
e-mail: manuel.guedes@tecnico.ulisboa.pt

A. Sayigh (ed.), *The Importance of Wood and Timber in Sustainable Buildings*, Innovative Renewable Energy, https://doi.org/10.1007/978-3-030-71700-1_6

production of qualified wood nor has the supply of national timber products contributed to encourage the use of wood.

The wood, cork, and pulp industries are the three major industrial sectors that depend on the forest, but the production of graded structural wood has never been significant in this sector. More than half the volume of pinewood is consumed for biomass, pellets, pulp, and panels. Sawmills, despite consuming most of the pinewood, concentrate their offer on low-quality wood products, specializing in low-value goods. In addition to the low demand for national pinewood for structures, this reality is also justified by the scarcity and high cost of pinewood from mature trees.

Although some authors [2] refer to the low quality of Portuguese timber for structural uses, the maritime pine ("Pinho-bravo," *Pinus pinaster*), one of the dominant national species, has the potential to be used for the manufacture of glued structural components, such as floor and roof beams (glued laminates). In addition to the maritime pine, also the Pyrenean oak ("Carvalho-negral," *Quercus pyrenaica*) and the abundant eucalyptus ("Eucalipto," *Eucalyptus globulus*) could be used for more specific solutions. The first, due to its high natural durability, would have a vocation for use in high-quality singular structural components, and the second could be used in small structural components, including roof structures with solid components. Mention should also be made to the cryptomeria ("Criptoméria," *Cryptomeria japonica*) wood, because it is abundant in the Azores, and the stone pine ("Pinheiro-manso," *Pinus pinea*) wood, which have been used in construction, deserving further study about its characteristics and uses.

Any potential investors in a qualification process of the Portuguese forest for the production of structural timber will face the risks of a small market and the constraints and uncertainties related to wildfires, pests (such as the pinewood nematode), and invasive diseases. Climate change is also an additional threat. In addition to the associated risk of forest desertification, pine and eucalyptus forests in Northern Portugal are expected to be replaced, in the future, by forests with typical species of the south of the country, such as the cork oak forests ("montados"). Additionally, global trade makes available wood products from countries that offer high-quality products (certified and classified products, from forests with selected species) with very competitive prices, which is hard to beat.

Climate

Historically, climate was one of the contextual factors that most conditioned the construction typologies. In addition to the availability of materials, climatic differences led to the development of specific construction systems in different regions. The generic relationship between climate and architecture allows to point out some typical and logical associations. Thus, in tropical areas, wooden structures and constructions have usually been associated with the importance of the roof for rain protection and with the open space (through post and beam structures) to optimize ventilation, and in hot and dry climate, wooden structures provide support for earthen or masonry walls, necessary for thermal inertia [3]. In cold regions, heavy wooden structures (usually using solid wood logs) were related with heat conservation strategies and closed-compact spatial organizations.

Portugal is located in the south-temperate climatic zone. In territories subject to this climate, there has been a traditional predominance of stone or brick masonry constructions. Wooden buildings would begin to appear more frequently in northern-temperate zones, although coexisting with masonry building [4]. This physical context, which has become cultural, can be pointed out as one of the main reasons why the tradition of wooden construction in Portugal is not comparable to the one of countries located to the south, in the equatorial and tropical regions, nor to the countries located to the north, in the areas of cold forests and mountainous regions.

Mainland Portugal includes two climatic zones: to the north the temperate zone of dry and moderate summers and to the south the temperate zone of dry and hot summers. In the "Bioclimatic Concepts for Buildings in Portugal," published by the Institute of Engineering Technology and Innovation [5]—INETI [6], bioclimatic strategies were proposed for the design of buildings according to these climatic zones. Essentially, the recommended strategies were common to all regions of the country, but their practical application would depend on the climate severity of each one. Regarding the winter conditions, the proposed strategies consisted of promoting solar gains, restricting conduction heat losses, and promoting strong thermal inertia. The measures to be adopted should focus on the application of insulation all over the building envelope and the use of heavy walls with external insulation, especially in some regions of the interior northeast. With regard to summer conditions, the proposed strategies pointed to the control of solar heat gains and conduction heat gains and to the promotion of ventilation, recommending also the use of strong inertia for almost all the national territory. The suggested measures would consist of shading glazed surfaces, isolation of building envelopes, transversal ventilation, and use of heavy walls with external insulation.

The abovementioned strategies and measures seem to favor systems that, unlike wood, use high-mass materials and are materialized by masonry walls, with high thermal inertia. However, based on the observation of winter conditions, considering the good thermal insulation characteristics of wood construction systems, in spite of their low thermal inertia, it can be said that these could have a good behavior in most of the national territory, mainly in the north. The lower severity of summers in that area, excluding some northwestern regions, would confirm this suitability. Based on the observation of the regions where the few examples of popular wooden architecture can be found, coastal areas and river basins could also be understood as naturally appropriate (justified by the moderating effect of water breezes) and some mountain and forest regions.

If kept in favorable conditions, wood has potentially unlimited durability. However, since those favorable conditions do not occur in nature, wood, as it is biodegradable, declines when subjected to determined living agents. Among these, the attack of wood by fungi, which occurs under certain conditions of humidity and temperature, is the one that causes a greater degree of degradation, usually occurring in countries with tropical or temperate climates (like Portugal). The absence of experienced technicians in Portugal can lead to architectural solutions that are not very durable due to the ignorance of the effect of the general conditions of precipitation and wind that expose buildings to water, causing its ingress and accumulation inside the components. Unawareness of other phenomena, such as the formation of condensation inside external walls and wooden roofs, will result in countless construction pathologies that may discourage any new attempt to use wood.

In addition to fungi, xylophagous insects pose an additional risk for wood construction. In Portugal, some of them are common [7]: the European house borer ("Caruncho grande," *Hylotrupes bajulus* L.), which attacks mainly coniferous species and in general only sapwood; the furniture woodworm ("Anobium," *Anobium punctatum*) associated with infestation of softwood and hardwood furniture; the powder-post beetle (*Lyctus*) that attacks mainly hardwoods, rich in starch; and the dry-wood termites (Isoptera: Kalotermitidae). This contextual risk requires that good design practices are known to guarantee the durability of buildings, not only from the point of view of the choice of wood species and their treatment but also through durability approaches based on a "durable design." Ignorance of these methodologies is sometimes responsible for pathologies that lead to false conclusions and prejudices regarding the use of wood.

Culture

Although the analysis of Portuguese vernacular architecture does not show a strong tradition of integral wood construction, its practice existed at a certain point. The interruption of this tradition is one of the factors that partly explain the current difficulty in accepting and adopting alternative wood solutions. In opposition to what happens in many other countries, Portugal cannot rely on the positive role of tradition.

The most important historical examples, with partial wooden structures, still existing, are the medieval buildings of Lisbon and Porto, the warehouses and dwellings of Aveiro, the eighteenth century buildings and "gaioleiro" structures in the downtowns of Lisbon and Vila Real de Santo António, and some buildings in other urban centers such as Lamego, Guimarães, Vila Real, and Chaves. However, most of these cases correspond to solutions in which wood is a partial structural solution and where coatings are made with other materials. Concerning integral wooden structures, mention should be made to smaller vernacular houses, generally associated with fishing villages, especially the ones known as "palheiros," on the central coast and the Tagus River Valley.

Structural wooden walls come with a wide variety of solutions in Portugal. Among them, three main types are well known: (1) wood covered with wooden planks, (2) wood with mortar filling, and (3) wood with masonry filling. The first solution is that of a wood frame that sometimes becomes a post and beam structure, coinciding with the solutions of the "palheiros" from the central coast and "Avieiros" along the Tagus River. The second solution is that of heavy wood frames, whose voids are filled with earth mortar, sometimes structured with straw, being observed in some urban buildings in the north of the country. The last solution is also the one of a heavy wood frame or a half-timbered frame but with the filling in masonry, identified as "taipa de rodízio" in Guimarães, assuming other designations according to regional variants ("frontal à galega," "enxaimel," and "gaiola pombalina"). It could also be considered a fourth type, for interior walls, covered with mortars ("tabique" or "taipa de fasquio"), consisting of a structure of planks ("tábuas costaneiras") arranged vertically, and to which a wooden slat is added where mortars are applied.

Due to its uniqueness and its relevant systematic characteristics, the "gaiola pombalina" ("Marquis of Pombal cage") is probably the most studied wooden construction system in Portugal. Its conception and adoption coincided with the reconstruction of Lisbon after the earthquake (1755), being invented as a construction system specially designed to offer safety conditions in the event of earthquake. The "gaiola pombalina" frame is basically built by a set of "posts" and "plates," the latter being connected by small wooden components, called "hands," to the walls. These elements are braced by X-shaped timber elements called Saint Andrew's crosses. These buildings are based on a three-dimensional-framed structure formed by walls, floors, and roof. There are three types of wall in this system: the external walls, in rubble stone masonry, mainly resisting gravitic loads; the interior timber-framed walls, named as "frontais," supporting floors and resisting earthquake horizontal/shear loading; and, finally, the simple partition walls, called "tabiques."

After the earthquake, Lisbon was rebuilt with high-quality buildings, but the construction between the nineteenth and twentieth centuries witnessed a new building type called "gaioleiro," corresponding to a degeneration of the "gaiola pombalina." The wooden structure for roofs and floors was a solution widely used throughout the national territory until much later, becoming sporadic as reinforced concrete was consolidated as a standard solution. Many of these solutions are part of the urban fabric of cities and towns in Portugal, offering opportunities for refurbishment interventions such as the one carried out by Tiago do Vale in the "Three Cusps Chalet" (Figs. 6.1 and 6.2), where there was the will to preserve the spatial and functional distribution, the wooden floors and ceiling structures, and the original staircase.

In the context of Portuguese rural architecture, the current use of wood is found mainly in floors and roofs, taking advantage of its characteristics, in terms of lightness and resistance to bending. The "Popular architecture in Portugal" survey, carried out between 1955 and 1960 [8], divided the country in different research regions. About zone 1 (from Minho to Mondego), it is said that in certain regions, the use of wood is very widespread, but it is also mentioned that there are no cases where its use is integral. The authors refer, however, that, taking into account the cost, ease of transport, and equipment, which does not require specialized labor, such as that of the bricklayer, wood would undoubtedly be the most suitable material to be applied, systematically, if it was not for its short duration. The greater the humidity of the soil, which is quite large in these regions of heavy rainfall, the

Fig. 6.1 The Three Cusps Chalet, Braga—Tiago do Vale (2013)—Photo: João Morgado

Fig. 6.2 The Three Cusps Chalet, Braga—Tiago do Vale (2013)—Photo: João Morgado

shorter the duration. The concern with the durability of structures subject to conditions of high humidity is thus a justification for not finding integral wood solutions, even where their use would be justified in view of other arguments. It is in zone 4 of the survey that references to coastal "palheiros" appear as wooden constructions on piles, tiled roofs, and generally one floor. The authors sustain that the wooden construction is promoted by the presence of the pine forest and, at the same time, wood works correctly in relation to the sandy soil and the humidity that the sea air brings.

Fig. 6.3 Wooden houses at Costa Nova. Conceptual reference for Centro Cultural da Costa Nova—ARX Portugal

Other authors [9] studied the Portuguese coast between Espinho and Praia de Vieira de Leiria, where the use of wood in the construction of houses was widespread at least from the end of the seventeenth century; see Fig. 6.3.

In their research, they state that, in most situations, these houses would assume characteristics of "stilt buildings" to prevent the invasion and accumulation of sands carried by the wind. The use of wood would be explained by several factors: the absence of stone and clay; the difficulty of transportation in the sands; the availability of wood in the pine forests planted along the dune; and, finally, its lightness in relation to other materials.

In the Madeira and the Azores islands, the "stone house" became the dominant model of rural housing. Nevertheless, wooden architecture is occasionally shown with great character. [10] Jose identifies two building types in which the use of wood is integral: the wooden and straw houses, also called huts ("cabanas, choupanas, palhotas"), and the wooden and stone houses with tiled roofs. In the first type, two subtypes can be found: the so-called "Santana da Madeira" houses, with a facade on the gable and three roof planes, with wooden structure and thatch finish, and another subtype, with vertical wooden walls and four roof planes, also roofed with thatch. The second type appears in the Azores islands (Faial and S. Miguel), presenting itself as a house in which a wooden elevated floor, with a tiled roof, sits on ground floor, built with stone masonry.

This building tradition proved to have little significance, both in terms of diffusion and visual impact. On the one hand, small wooden buildings had no relevant expression in terms of occurrence in the territory, and on the other hand, urban structures of greater importance hid their wooden components (floors, roofs, and partition walls) inside their external walls. It should be noted, however, that there are notable exceptions which have served as a reference for the choice of wood as a constructive solution. Atelier ARX Portugal, for instance, describes the project for the Socio-Cultural Center of Costa Nova (Figs. 6.4 and 6.5) in this way: "The typology of construction and original urbanity of Costa Nova constitutes one of the main conceptual premises of the project, which it seeks to reinterpret, in a clearly contemporary approach, the way of looking at the place and building, ancestral and characteristic of Costa Nova. That is why, all the construction that is made of wood, is based on a semi-buried network of reinforced concrete foundations, in order to stabilize the building that will 'float' on the dune, just like the constructions of the original haystacks" [11].

Also, about the Dovecote-Granary project (Figs. 6.6 and 6.7), Tiago do Vale, the author, writes, "Originally built in the late XIX century, its starting point were two traditional northern Portugal maize granaries standing over granite bases. (…) Times changed, though: there's no farming on the property anymore, so the rebuilt Dovecote-Granary will not serve its original functions in the foreseeable future. It won't have a specific use either: it will be what the nature of the space lends itself to be. (…) The result is an element by element rebuild of the Dovecote-Granary, with an intricate redesign of all the subtle carpentry details and a limited set of surgical interventions that will allow for its safe and renewed use. (…) The Dovecote-Granary is now a sanctuary among the tree canopies, an iconic shape in the rural landscape of the Minho region, and the experience of the dancing leaf shadows, the gentle crossing breeze and the birds chirping in a late summer afternoon fully defines its new purpose, function and use" [12].

Fig. 6.4 Socio-Cultural Center of Costa Nova, Ílhavo—ARX Portugal (2015)—Photo: Fernando Guerra | FG + SG

Fig. 6.5 Socio-Cultural Center of Costa Nova, Ílhavo—ARX Portugal (2015)—Photo: Fernando Guerra I FG + SG

The Portuguese Architectural-Constructive Way

Among the generation of architects that, in the 1980s and 1990s, elevated contemporary Portuguese architecture to an international level (Álvaro Siza, Souto de Moura, Carrilho da Graça, Gonçalo Byrne, etc.), none of them were interested in wood construction. The design approaches that these architects were adopting since the 1970s constituted a base of influence for the following generations. The idea of a "Porto school of architecture" or of a "Portuguese way" was then manifested by the purity of forms and the contrast between light and shadow, whose principles are very compatible with the technology of masonry, reinforced concrete, and plastered and painted white coatings, sometimes with limestone or granite basements, floors, and fittings. Since structural wood has never been a technology explored by these reference authors, most students of architecture and young Portuguese architects have also never found motivation to research in the "unknown" field of wood construction.

Fig. 6.6 The Dovecote-Granary, Ponte de Lima—Tiago do Vale (2016)—Photo: João Morgado

The interest, curiosity, and experiences in wood construction were initiated in Portugal by small companies of prefabricated single-family homes. The solutions offered by these firms were often based on traditional typologies, imported from other countries, or were simply very basic solutions, where low cost and ease of assembly were the main objectives. Architects were left out of this market, and for many years, wooden construction was associated with a specific market niche and poor-quality solutions (although there were exceptions), temporary, or with a design aesthetics seen as inappropriate to the national context. Some of the prejudices regarding wood construction were also related, not so much to the solutions of these companies, but more to two other aspects that have given wood a reputation of a poor, temporary, and nondurable material. Such was the case of the neighborhoods of illegal shacks on the outskirts of large cities, built in part with wooden elements, and the temporary school buildings, also built with wood, after the revolution of 25 April 1974.

Fig. 6.7 The Dovecote-Granary, Ponte de Lima—Tiago do Vale (2016)—Photo: João Morgado

Carlos Castanheira is one of the Portuguese architects who most consistently bet on wood as a building material and also one of the great promoters of its architectural possibilities. The solutions he initially adopted were often about the combination of wood with other materials. His wooden house in Quinta do Buraco, Oliveira de Azeméis (Figs. 6.8 and 6.9), built in 2001, for example, rises from a concrete platform. As the author states, "From the concrete base upwards, all the pillars, beams, linings and door and window frames are made of wood, produced by excellent carpenters, who fortunately can still be found. (…) The exterior covering is constituted by wonderful manufactured copper, glass and slate on the floor areas adjacent to the interior floors" [13].

Also, in this same spirit, the Avenal house, built in 2004 (Figs. 6.10 and 6.11), was built with a Riga wooden structure assembled on top of concrete floor slabs, coexisting with some masonry walls. "The exterior is constituted by slate walls, and the ashlar masonry work and wall-corners are reinforced with granite. The front of

Fig. 6.8 Quinta do Buraco house, Oliveira de Azeméis—Carlos Castanheira (2001)—Photo: Fernando Guerra | FG + SG

Fig. 6.9 Quinta do Buraco house, Oliveira de Azeméis—Carlos Castanheira (2001)—Photo Fernando Guerra | FG + SG

the building, looking South, is constituted entirely of wood and glass" [14]. The Adpropeixe house (Figs. 6.12 and 6.13) in Terras de Bouro, built in 2008, one of the most famous, was taken one step further. "Concrete footings (the only concrete used in the building). Steel connectors. Structural frame, internal floors and ceilings in Scots Pine, external flooring and ceilings in Cumaru Brazillian Teak. External finishes, walls and roof in copper. Glass in timber frames. Posts with fixings in copper and a wire mesh of polypropylene nylon form the fencing" [15].

Fig. 6.10 Avenal house, Oliveira de Azeméis—Carlos Castanheira (2004)—Photo: Fernando Guerra | FG + SG

Fig. 6.11 Avenal house, Oliveira de Azeméis—Carlos Castanheira (2004)—Photo: Fernando Guerra | FG + SG

Fig. 6.12 Adpropeixe house, Terras de Bouro—Carlos Castanheira (2008)—Photo: Fernando Guerra | FG + SG

Fig. 6.13 Adpropeixe house, Terras de Bouro—Carlos Castanheira (2008)—Photo: Fernando Guerra | FG + SG

Carlos Castanheira considers that two of the advantages of building with wood are the need to foresee all problems before the construction phase and the minimization of waste on the construction site. The materiality of the wood fascinates him: "I've used timber because I like to! I like the smell; the touch, the natural colors, the veins, the knots, the mullion and the transom, the post and the beam, the dowel and the tenon; the chisel and the gauge… I like… carpenters." The process turns out to be an excuse for thinking about construction: "Just as if you were building a piece of furniture, a very large one, building in timber forces one to foresee, to think, to organize, to organize the thinking; because all of the infrastructure has to be considered in advance and installed before and during the construction process, which itself becomes an assembly process, where assembly is a rational activity" [16]. Paradoxically, these high-quality projects did not have served to influence a large number of Portuguese architects, nor to persuade the traditional construction market, conditioned by the low budgets of the owners and their low-quality culture. Works of this quality remain an exception.

Lack of Incentives and Initiative

Since it is difficult to introduce new materials in markets where reinforced concrete and steel structures were made dominant by their efficiency and widespread know-how, some countries, convinced of the benefits of wood, turned to specific policies to encourage its use in construction: information and promotion of voluntary actions, introduction of environmental standards, legislation to encourage the use of a certain proportion of wood in buildings, updating technical and structural standards (earthquake, fire safety, maximum heights, etc.), and public procurement policies. In Portugal, this type of incentives, aimed at promoting the use of wood in structures, does not exist.

National institutions, even those involved in the forest and wood sector, while not ignoring that timber construction can be a beneficial solution within the framework of a sustainable development strategy, seem to ignore the importance of structural solutions. The 2006 Forestry National Strategy—ENF [17], for example, states that in addition to envisioning the alteration and qualification of forestry practices and certification, it is important to set the "mobilization of agents for the use of forest products, such as designers and architects." The document mentions the importance of the construction sector as a potential user of products derived from wood and cork and its relevant role in the creation of long-lasting carbon storage. It is also stated that "their use, for example, should be promoted through mechanisms of positive discrimination in construction activities and remodelling of public buildings." Until today, however, concrete steps in this direction were not taken.

Another factor that seems to contribute to the weak dynamics of this type of construction is the lack of joint efforts and organization, by the various companies in the wood sector, toward the specific goal of promoting wood products. Contrary to what happens in many other countries, there is no associations formed specifically for this purpose, influencing and lobbying the authorities to take concrete measures in favor of wood construction. Associations and campaigns such as "Wood for Good," "Canadian Wood Council," "Wood Works," "Wood Solutions," "Swedish Wood," "American Softwoods," and "Comité National pour le Développement du Bois" try to convince the authorities (governments) to support the use of wood: Reducing VAT on wood products, spreading public information on wood as an eco-material, promoting the use of wood in public buildings, defining methods for assessing carbon storage and the cycle of material's life, etc.

The Portuguese Association of Wood and Furniture Industries (AIMMP) "is the only business association in the sector with a national scope and with a global wood sector perspective, representing all forest-based industries, except cellulose, paper and cork" [18]. This association, in spite of promoting actions and projects in the context of the wood industries, is not aimed at promoting wooden structures. The projects that the association has underway are betting on internationalization, company innovation, training, and forest management, with a very strong focus on the furniture industry.

In 2010, the National Strategy for Adaptation to Climate Change (ENAAC) was approved [19]. In 2013. a progress report was done, mentioning new strategies including "Betting on sustainable construction, R&D and eco-innovation, through: (…) Valuing innovative projects that are a factor of differentiation and sustainability; (…) Use of building materials adapted to the worsening of risks, for example for heat waves (based on cost-benefit analysis)" [20]. Despite references that indirectly refer to construction materials, wooden construction and its potential for storage and carbon sequestration are completely ignored in this text.

6.2 In Favor of Wood

Emergence of Sustainability

The evidence of climate change and the emergence of sustainability as a development objective have drawn attention to the role that the use of wood can play as a building material. Although in Portugal there is still not a dynamic toward the substitution of traditional materials by wood, the context for this to happen in the near future seems to be being launched.

The abovementioned 2006 Forestry National Strategy (ENF) was revised in 2015, after being subjected to an evaluation process that tried to integrate the new developments of the European Union Forestry Strategy, the European Union Biodiversity Strategy, and the Strategy for Smart, Sustainable and Inclusive Growth. In addition, the National Strategy for Adaptation to Climate Change (ENAAC) and the National Action Program to Combat Desertification (PANCD) were also taken into account [21].

In the strategic operationalization matrix of the Forestry National Strategy—ENF, the strategic objective "D," concerning the internationalization and increasing the value of products, proposes actions of raising the awareness for the use of forest products and promoting the use of forest products through public procurement. It should be noted that one of the measures included in the objectives of the Azores forestry strategy is to implement the CE marking (European Conformity) process for *Cryptomeria japonica* wood and to promote R&D studies in the field of forest products' technology. The idea is to work on a Cryptomeria Wood Technical Sheet, according to European norms, with a mechanical grading that allows to assign the quality class for the use of its timber in construction works, structures, or other uses.

The Forestry National Strategy (ENF) also states that the public sector can play an important role in promoting the use of wood and cork through positive discrimination in the use of forest products, for example, in purchases for the construction or rehabilitation of public buildings. Also, the use of wood may be promoted by prioritizing of products of forest origin in "public procurement," namely, within the scope of the National Strategy for Ecological Public Purchases—ENCPE [22]. Within the scope of this strategy, the "Public Procurement Guide" states that the state, as a contracting entity, has the right to insist that the product it intends to purchase shall be made from a specific material or shall incorporate a certain percentage of materials/components, recycled or reused. Although wood is never referred to as a material on which positive discrimination can be made, wood will undoubtedly be one of the materials of choice to be favored in public procurement.

The idea of sustainability is also recurring by Portuguese architects who promote the use of wood in the justifications given for choosing this material. In addition to the aesthetic attributes associated with wood, aspects of reducing embodied energy, reducing greenhouse gas emissions, carbon sequestration, and other environmental impacts (reduction of water consumption, construction waste, etc.) are often mentioned as determinants of choice. On this subject, Carlos Castanheira says, "There is also the advantage that the rough work becomes the finished work, the exterior so often crosses over to the interior, without involving obsessive problems of thermal bridging; the beam is mistaken for the window frame and meanwhile that frame acts as a beam; the ceiling provides the flooring and the floor, the ceiling; simple solutions, if they are accepted, tolerated" [16]. Appleton & Domingos, comment on some of their projects: "Treehouse (Fig. 6.22) is based on sustainability and modularity. The houses grow according to the needs of families or individuals, as well as the new branches of a tree" [23]; "Rehabilitating means, as much as possible, the use of traditional, natural materials (wood, stone, sand and lime), as opposed to the use of artificial industrial materials such as cement, steel, aluminium, pvc and other polymeric materials, etc." [24]. Less mentioned are the benefits in terms of the advantages of choosing wood in the context of sustainability certifications. Probably in Portugal, the idea of certifying a project or a building is still seen by project owners as a cost and not as a benefit.

Sometimes, pragmatism and the need to build quickly combine with sustainability goals. This is how the use of wood at the Redbridge School in Lisbon was justified (Figs. 6.14 and 6.15) by the ARX Portugal office: "Wood was selected from the beginning as its main structural and finishing material, due to the pedagogical message inherent to its low impact in nature, the welcoming environment and the speed of construction that it provides—we only had one year to think and build this school" [25].

Probably, the information and awareness campaigns in favor of wood should be more targeted at the owners. As Carlos Castanheira says, "A good client is the most important element for a construction work" [13]. Several national projects

Fig. 6.14 Redbridge School, Lisbon—ARX Portugal (2018)—Photo: Fernando Guerra I FG + SG

incorporated wood because it was included in the project's brief or requirements, directly by mentioning the material or indirectly through the sustainability objectives demanded by the owners.

Some of Carlos Castanheira's choices were fermented by the client. In the Adpropeixe house (Figs. 6.12 and 6.13), "Easier to reach by boat than by land or hill, the place is unique and the answer to the specific order would have to be unique; a wooden house" [15]. At Revigrés offices (Figs. 6.16 and 6.17), "The new Revigrés offices have very clear parameters set out by their future users: proximity to the new production area, functionality, sustainability and if possible identity and harmony. (…) Containers that are designed and made to measure, used and then later, are re-used within a logic of functionality and responsible sustainability. (…) This project is sustainable because there is a strong will to be sustainable and because, this willingness itself is sustained by reasons which are beyond those of mere necessity. (…) The containers have a softwood structure; the walls, floors and ceilings are made out of chipboard panels."

Fig. 6.15 Redbridge School, Lisbon—ARX Portugal (2018)—Photo: Fernando Guerra | FG + SG

"The thermal and acoustic insulation is provided by black expanded cork agglomerate." [26]. "The Equestrian Center located at Cabo do Mundo, Leça da Palmeira (Figs. 6.18 and 6.19), is inhabited by horses and working there, people who love horses. The challenge, or brief, required the use of timber in the structure and also in the partitions, walls and ceilings. A timber stable building with two enclosed arenas, a cellar and a social building, also in timber. Just as in any other project, it is necessary to listen to the clients and hear their requests, requirements and aspirations" [27].

Evolution of Construction Products

The reference to some of the contextual factors considered to be constraints on wooden construction is meaningless if one takes into account the reality of structural wood construction systems and contemporary construction materials. Most

Fig. 6.16 Revigrés offices, Águeda—Carlos Castanheira (2013)—Photo: Fernando Guerra | FG + SG

Fig. 6.17 Revigrés offices, Águeda—Carlos Castanheira (2013)—Photo: Fernando Guerra | FG + SG

Fig. 6.18 Equestrian Center, *Leça da Palmeira—Carlos Castanheira (2017)—Photo: Fernando Guerra | FG + SG*

Fig. 6.19 Equestrian Center, Leça da Palmeira—Carlos Castanheira (2017)—Photo: Fernando Guerra | FG + SG

building systems include walls, roofs, and floors, made up of several layers, with the wooden structural components often being protected by other materials that can work as thermal insulation or as thermal mass. The lightness of wood and its reduced mass can then be compensated with suitable construction solutions and by adding other components. Thus, if the universe of "wooden construction" is understood in a rigorous way, it must not be considered as inappropriate for a country like Portugal, despite its temperate climate.

Fig. 6.20 Municipal swimming pool—Sport Park of Fróis, Almada—J.A. Arquitectos (2011)—Photo: Luis Morgado

The promotion of thermal mass is a constant in the strategies recommended for Portugal, both for summer and winter. Thermal mass can be enhanced in wooden constructions, whether post and beam, wood frame, or panels, through cladding or by additional elements (ventilated facades, floor slabs, fireplace cores, stairs, single masonry walls, etc.). For the same purpose, heavier wood systems, such as CLT panels (Fig. 6.20), or thermally insulated logs can also be chosen instead of traditional light systems (wood frames). Besides thermal mass, attention must also be paid to the recommended measures to mitigate solar gains in summer (through the protection of openings, adequate glazing, and roof projections). In this context, it is important to have a notion of the diversity of wood systems and products, so that architects can make the right choices for each situation.

In addition to a small number of new companies that started to invest in wooden construction, some of the oldest wooden houses and carpentry companies have adapted to the new market challenges and have evolved to offer a wider range of products, including the most innovative laminates and panels. These companies presently have a portfolio of solutions designed by architects and offer services aimed at developing more complex projects. The increase in the quantity of wooden construction occurred within a short time, leading to the accumulation of projects in the few offices that showed competence and interest in a material that required different methods and routines. Thus, in addition to the technical offices of construction companies, some engineering offices are recognized today for their specialization in the design of wooden structures.

European Codes

Due to the weak dynamics of wooden structures in Portugal, national regulations were never developed to the point of serving as a guarantee and a reference for designers to comfortably consider their use. Portugal's membership in the European Union solved this problem since the structural design of wooden structures started to refer to "Eurocode 5—Design of timber structures," which takes into account the characteristics of the materials and components used (mechanical properties depending on strength classes). In addition, the actions to be considered in the calculation (dead loads, live loads, environmental loads, and other actions) may be referenced to the Portuguese Code for Safety and Actions for Building and Bridge Structures [28] or also to "Eurocode 1—Actions on structures" and "Eurocode 8—Design of structures for earthquake resistance."

The classification of wood components is one of the fundamental aspects to allow for the raw material to be used in structural elements. In Portugal, some consider that the grading standards were inadequate and that sawmills were unable to classify the components they produced. However, Portugal benefits from the standardization of the requirements for classification in the European Union defined by the EN 14081 standard, with the strength classes established in the EN 338 standard: 12 classes for conifers and poplar (C14–C50) and 6 classes for hardwoods. The association between the visual classification and the strength class system is made through the standard EN 1912

It is also important to mention that several Portuguese institutions are prepared to give scientific support to the implementation of wood in construction namely, the National Civil Engineering Laboratory—LNEC, which throughout its history has been providing support in terms of research, certification, and homologation of products, as requested by wooden manufacturing and construction companies. As for universities, there is a notable growing interest in wood construction in both engineering and architecture faculties, which, in recent years, have developed several research works in this area.

New Architectural Visions

Besides companies of wooden house construction, the initial impetus for some dynamics of wooden construction in Portugal was given, at the end of the last century, by the various sports pavilions that were built incorporating laminated beams of glued wood (glulam) in large spans. Perhaps the most paradigmatic work was the "Pavilhão Atlântico" (Fig. 6.21), currently called Altice Arena, completed in 1997. Designed by Architect Regino Cruz (in collaboration with Skidmore, Owings & Merrill), it was also an innovative project at the international level because it was one of the first large structures to be calculated according to Eurocode 5. Composed of transversal frames (composed of main and secondary beams and connecting arches), ranging between 52 and 115 m in length, it has an area of 42,600 m^2 and

Fig. 6.21 Pavilhão Atlântico, Lisboa—Regino Cruz—Foto "Pavilhão Atlântico"

capacity for 20,000 spectators. It was an Energy Comfort 2000 Building under the European Commission's Thermie Project. Based on the success of this case, countless buildings followed, with large-span halls, taking advantage of the aesthetic and structural performance of wood.

Subsequently, several architects began to explore solutions with wooden structures, in typologies hitherto dominated almost exclusively by the construction companies of wooden houses. In addition to Carlos Castanheira, one of the most interesting examples is that of the Arquiporto studio, together with the company Modular System, which greatly contributed to changing the image of wooden construction in Portugal, associating it with contemporary, clean, and well-detailed architectural forms. Representing this innovative attitude, in Vieira do Minho, Arquiporto [29] has built in 2001 a house in Caniçada, Gerês, with glued laminated wooden structure where the company's philosophy was put into practice through the use of raw materials, without coating or cladding. According to the authors, technological issues, as acoustics, wood torsion, and connections between wood beams and stone walls, were as important as the stylistic issues [30].

Other examples where wood was used in a creative way emerged, both with the aim of building unique solutions, based on traditional systems, on one side, and with the ambition to renew the modernist idea of industrialized series production, on the other side. As an example of the first case, one can mention the intervention of the architects in Aires Mateus in Comporta [31]. They started from two constructions in wood and two in masonry to achieve an elementary architecture adopting the language of the construction process of vernacular thatched huts of the south coast of

Fig. 6.22 Treehouse exhibition, Azambuja—Appleton & Domingos (2011)—Photo: Luis Morgado

Portugal. In the second case, the work of the duo Appleton & Domingos is remarkable, in collaboration with the company Jular [32], with a set of well-designed residential modular solutions. The "Treehouse" (Fig. 6.22), for example, was a project developed through a modularity strategy that aimed to offer the possibility of expansion by adding modules. Another significant example of modular projects, with an innovative solution in terms of design, resulted from the work carried out by the Atelier MIMA architects [33]. With a very simple approach, they proposed solutions that distanced themselves from the traditional language of wood.

The possibilities of the wood material to fit into landscapes, where a minimal human intervention is necessary, were decisive in the choice of several architects. In the case of the Environmental Interpretation Center EVOA (Figs. 6.23 and 6.24), designed by the Maisr Arquitectos studio: "The skin of the building reflects the intention of integration with the landscape, using the wood like a natural element, with an expression that refers us to the image of reeds and palisades. It is anticipated that the exposure of the wood to the environment contributes to modifying his natural tone and becomes similar to the color of the surroundings environment" [34]. In a similar way, the Treetop walk of Serralves, Porto (Figs. 6.25 and 6.26), is described by Carlos Castanheira as follows: "We are walking on timber, almost everything in timber. Structure, walkway, balustrades become one with the trunks and branches. In time they will become ever more alike and the walkway will eventually be as if it had always been there. Like a growing tree that asserts its place naturally" [16].

After the "founding projects" of this new technological approach, many other wooden architecture solutions, innovative in the national context, were carried out

Fig. 6.23 Environmental Interpretation Center EVOA, Vila Franca de Xira—Maisr Arquitectos (2012)—Photo: Filipa Miguel Ferreira

Fig. 6.24 Environmental Interpretation Center EVOA, Vila Franca de Xira—Maisr Arquitectos (2012)—Photo: Filipa Miguel Ferreira

Fig. 6.25 Treetop walk, Porto—Carlos Castanheira (2019)—Photo: Fernando Guerra | FG + SG

Fig. 6.26 Treetop walk, Porto—Carlos Castanheira (2019)—Photo: Fernando Guerra | FG + SG

by Portuguese architects. In addition to projects of single buildings, also, urban and touristic developments were a modality that stood out as a strong field of application for wooden construction. Some of the best-known examples are the Cocoon Eco Design Lodges project at Comporta [35], consisting of 30 wooden houses promoted by Modular System; the project Zmar Eco Campo Resort & Spa in Odemira, with 50 wooden houses, built by Jular [36]; and the Pedras Salgadas Eco-Resort [37] and Pestana Tróia Eco-Resort [38].

Only after the beginning of the twenty-first century was wood construction began to be understood by Portuguese architects as having an exciting architectural potential. The most recent projects with wooden structures that show its use, in addition to the environmental argument, represent a possibility of renewing Portuguese architecture, free from the references that marked the end of the twentieth century. New architecture trends are, probably, more focused on the values of landscape and place and less on the idea of creating striking works and new sculptural references. This is how Atelier [A] Ainda Arquitectura describes their objectives for the Mountain House project (Figs. 6.27 and 6.28) in Louredo, Vieira do Minho: "our work as architects was aimed at minimizing the impact of the house against the forest, and making the final design seem as effortless as possible, investing great part of the work in precisely effacing this effort. (…) How to take advantage of the

Fig. 6.27 Mountain House, Vieira do Minho—[A] Ainda Arquitectura (2010–2017)—Photo: Fernando Guerra | FG + SG

Fig. 6.28 Mountain House, Vieira do Minho—[A] Ainda Arquitectura (2010–2017)—Photo: Fernando Guerra I FG + SG

natural beauty of the emplacement without destroying the thing you love? I suppose the relevance of this design is not really about the house, but about the site. About making the house disappear" [39].

Policies

In the scenario formerly described, the state and the Portuguese institutions are not concerned with wood construction. Perhaps a sign that the panorama is changing in Portugal is given by the recent changes in the structure of the ministries of the Government of Portugal. The responsibility for the forests was transferred from the Ministry of Agriculture to the Ministry of Environment and Climate Action. On the other hand, the Platform for Monitoring Relations in the Forestry Sectors (PARF)

[40] will have a new dynamic, integrating environment and economy, through the Forest and Nature Conservation Institute (ICNF) and the Directorate-General for Economic Activities—DGAE. Its objectives now consist, among others, of the data analysis about forest resources, the elaboration of studies from the perspective of supply and demand for forest products, collaboration with the scientific community, and the elaboration of proposals for the regulation of the sector.

Part of the success of timber construction depends not only on governmental institutional policies and organizations that currently support the interests of the forest and the timber industries, like the Portuguese Association of Wood and Furniture Industries (AIMMP), the Forest and Nature Conservation Institute—ICNF, and the Pinus Center—CP. It is also influenced by the interest of universities, architects, engineers, and public opinion, through the clients, who are ultimately responsible for the final decision.

In the absence of direct incentives to build with wood structures, for the time being, probably one of the most relevant actions for its promotion has been the National Award for Architecture in Wood [41]. The prize is promoted by the Portuguese Association of Wood and Furniture Industries—AIMMP, which chairs the organization, together with the Portuguese Architects Association (OA) and the Portuguese Confederation of Construction and Real Estate—CPCI. It is not probably a very effective action toward changing the minds of the general public, but at least, it is the type of program that can have a positive and dynamic impact on the attitude of national designers.

The "Quinta do Orgal" (Fig. 6.29), designed by Menos é Mais Arquitectos Associados, was the winner of the fifth edition (2019) of the National Award for Architecture in Wood—PNAM [42].

Fig. 6.29 Quinta do Orgal, Vila Nova de Foz Côa—Menos é Mais Arquitectos Associados (2018)—Photo: José Campos

Once again, the importance of the program was a fundamental factor in the choice of wood: “The request is that of a small equipment for agro-tourism built in a wood modular system. Fifteen modules with the dimensions of 6.60 × 3.30 m, generate the container to the necessary spaces of functionality and interior comfort. The exterior spaces, in the shape of balconies, long terraces, guarantee the accesses and circulations, as well as the protection against solar rays and weatherproof. Underneath, sustaining all the wood construction, there is a metallic frame that rests in two pillars that overcome a span of 13 m in the centre and two symmetrical cantilevers of 6.60 m on each side. These pillars, with habitable space inside, shelter all the technical zones and necessary services” [43]. The vitality of wood construction and its potential to be boosted were demonstrated by the diversity of typologies and the quality of the solutions competing in this fifth edition of the award: hotels, museums of existing structures, single-family houses, a small pavilion, rehabilitation of interior spaces, a school, a chapel, and a restaurant.

6.3 Why Don’t We Build in Wood?

Culture is one of the most frequent justifications found to answer this question by most architects of a large group that were contacted. Tiago do Vale considers *that* “Part of the rejection of wood as a building material stemmed from the Modern Movement and its post-war motivations: it didn’t fit with the discourse of construction as an industrial operation, and it represented a world of craftsmanship and detail that collided with a new world of abstraction and formal purity. It was a memory from a past that didn’t matter anymore. Those motivations are long gone but its effects lasting. Architects are more willing to work with materials they have experience with (and that average builders can satisfactorily execute). Promoters equally favor the materials they know best. Unfortunately, wood became a bit of an unknown space outside the intervening’s comfort zone.” *Information* “and” *example* “are the tools to change it” [44]. The Maisr Arquitectos studio also states that “it is mainly a cultural issue, since our construction has essentially used masonry. We started with adobe, with mud, we went through stone masonry and we are still in the phase of brick masonry combined with a reinforced concrete structure” [45].

According to João Appleton, the historic replacement of wood by reinforced concrete and steel was the result of a double myth: wood came to be considered as a precarious material, of reduced durability, and as structurally limited, while concrete was overvalued in all its qualities [46]. Carlos Castanheira answers this question by pointing out the inertia of rooted habits: “Concrete, the so-called eternal material, the lightweight slab, the guarantee of never again having any problems; the hollow brick, a very Portuguese material that has not evolved at all over the last century; rough and fine renders; cover-all paint, (…) and many, many other things, generated a building culture that allowed no space for thinking, simply ‘this is how we always do it, as we always have done’”; “The challenge of not always doing things this way, of not wanting to always do it this way, is no easy task! Engineers with their structural calculation software; clients and their wives; builders [saying]: if it was done the usual way, it would be a lot cheaper" [16]. But sometimes, as the architect himself has demonstrated, the best way to convince clients is to set a precedent: "That is how my own house is: concrete for everything under ground, above timber, protected from the weather by copper sheeting. (…) Other houses followed, after people seeing mine, and I believe like Saint Thomas, that more will follow… because I like it! And the clients do too!" [16].

The idea that we are facing a multidimensional problem is expressed by [A] Ainda Arquitectura: "In the Portuguese context, wood is not used often mainly because of the cost of its construction, which is also a result of the material availability in the local context, which in turn results in lack of tradition of construction in wood. It is also, as the three little pigs tale exemplifies, the general perception that it is not Value for Money, as it is not as robust as masonry construction, and it is high maintenance where regular building maintenance is an issue" [47].

Regino Cruz answers the question "Why are architects and builders not using wood in their buildings?" criticizing its formulation and stating the need to build more with wood: "Indeed they are, though not to the extent they could be" [48].

6.4 Conclusion

In Portugal, in general, there are still the two questions available: one for using wood and another is against wood in buildings. Could the potential of wood in terms of sustainability, comfort, aesthetics, and performance stand in opposition to traditional fears in terms of durability, safety, maintenance, and costs? We hope that reason one may prevail in due course. The paper shows beyond doubt that wood has a prominent role to play in Portugal. It has been in use for many centuries, and some innovative architects are encouraging its use. Subsequently, several architects began to explore solutions with wooden structures, in typologies hitherto dominated almost exclusively by the construction companies of wooden houses. In addition to Carlos Castanheira, one of the most interesting examples is that of the Arquiporto studio, together with the company Modular System, which greatly contributed to changing the image of wooden construction in Portugal.

Part of the success of timber construction depends not only on governmental institutional policies and organizations that currently support the interests of the forest and the timber industries, like the Portuguese Association of Wood and Furniture Industries (AIMMP), the Forest and Nature Conservation Institute—ICNF, and the Pinus Center—CP. It is also influenced by the interest of universities, architects, engineers, and public opinion, through the clients, who are ultimately responsible for the final decision.

References

1. INE. (2020). *Land use and land cover statistics*. Lisboa: Instituto Nacional de Estatística. Retrieved from https://www.ine.pt/ngt_server/attachfileu.jsp?look_parentBoui=439849394&att_display=n&att_download=y.
2. Sanz, F., et al. (2007). *Aplicações industriais do Pinho bravo*. Porto: AIMMP.
3. Correia Guedes, M. (2019). Bioclimatic project: General guidelines. In *Bioclimatic architecture in warm climates: A guide for best practices in Africa*. Cham: Springer. ISBN 978-3-030-12035-1.
4. Correia Guedes, M., Simões, R. N., Cabral, I., Barros, F. C., Carlos, G., Correia, M., & Marques, B. (2019). Vernacular architecture in Portugal: Regional variations. In *Sustainable vernacular architecture*. Cham: Springer. https://doi.org/10.1007/978-3-030-06185-2. ISBN: 9783030061845.
5. Ferreira, J. (2019). Masonry. In *Bioclimatic architecture in warm climates: A guide for best practices in Africa*. Cham: Springer. ISBN 978-3-030-12035-1.
6. Gonçalves, H., & Graça, J. M. (2004). *Conceitos bioclimáticos para os edifícios em Portugal*. Lisboa: INETI.
7. Machado, J. S., Dias, A., Cruz, H., & Custódio, J. E. (2009). *Avaliação, conservação e reforço de estruturas de madeira*. Lisboa: Verlag Dashöfer.
8. Keil Amaral, F., & AAVV. (1988). *Arquitectura popular em Portugal - 3 Volumes*. Lisboa: Associação dos Arquitectos Portugueses.
9. Veiga de Oliveira, E., & Galhano, F. (1964). *Palheiros do litoral central Português*. Lisboa: Instituto de Alta Cultura - Centro de Estudos de Etnologia Peninsular.
10. Fernandes, J. M. (1996). *Cidades e casas da Macaronésia*. Porto: FAUP Publicacões.
11. ARX Portugal. (2020). Centro Sócio-Cultural da Costa Nova. Text provided by the authors.
12. do Vale, T. (2020). The dovecote-granary—A peaceful retreat among the treetops. Text provided by the author.
13. Castanheira, C. (2001). House Quinta do Buraco III. Text provided by the author.
14. Castanheira, C. (2005). Avenal House. Text provided by the author.
15. Castanheira, C. (2008). Adpropeixe House. Text provided by the author.
16. Castanheira, C. (2001). Building with timber—Because I like to!. Text provided by the author.
17. Portugal. (2006). *Estratégia Nacional para as Florestas* (Resolução do Conselho de Ministros no. 114/2006).
18. AIMMP. (2020). *Associação das Indústrias de Madeira e Mobiliário de Portugal*. Retrieved from https://www.ine.pt/ngt_server/attachfileu.jsp?look_parentBoui=439849394&att_display=n&att_download=y.
19. Portugal. (2010). *Estratégia Nacional de Adaptação às Alterações Climáticas* (Resolução do Conselho de Ministros no. 24/2010).
20. APA. (2013). *Relatório de Progresso da Estratégia Nacional de Adaptação às Alterações Climáticas*. Agência Portuguesa do Ambiente. Retrieved from https://www.apambiente.pt/_zdata/Politicas/AlteracoesClimaticas/Adaptacao/ENAAC/RelatProgresso/Relat_Progresso.pdf.
21. Portugal. (2015). *Estratégia Nacional para as Florestas (Update)* (Resolução do Conselho de Ministros no. 6-B/2015).
22. Portugal. (2016). *Estratégia Nacional para as Compras Públicas Ecológicas 2020* (Resolução do Conselho de Ministros no. 38/2016).
23. Appleton & Domingos. (2012). Treehouse Riga/Appleton & Domingos. *Archdaily*. ISSN 0719-8884. Retrieved from https://www.archdaily.com/269096/treehouse-riga-appleton-domingos.
24. Appleton & Domingos. (2010). Reabilitação de edifícios antigos e sustentabilidade. VI ENEEC – Encontro Nacional de Estudantes de Engenharia Civil. *Évora*. Retrieved from http://www.neecue.uevora.pt/Documentos/VI-ENEEC/ENEEC%20reabilita%C3%A7%C3%A3o/Jo%C3%A3o%20Appleton%20-%206.%C2%BA%20Encontro%20Nacional%20de%20Estudantes%20engenharia%20Civil.pdf.

25. ARX Portugal. (2020). Redbridge. Text provided by the authors.
26. Castanheira, C. (2012). Containers made in wood. Text provided by the author.
27. Castanheira, C. (2020). Equestrian Centre—Cabo do Mundo. Text provided by the author.
28. Portugal. (1983). *Regulamento de Segurança e Acções para Estruturas de Edifícios e Pontes* (Decreto-Lei no. 235/83, de 31 de Maio).
29. Modular System. (2020). Casa na caniçada #01. Retrieved from http://www.modular-system.com/pt/projectos/casa-na-canicada-01/.
30. Neves, J. M. (2007). *Casas Ibéricas*. Casal de Cambra: Caleidoscópio.
31. Aires Mateus. (2011). Casa na Areia. *ArchDaily*. ISSN 0719-8884. Retrieved from https://www.archdaily.com/119742/casa-na-areia-aires-mateus.
32. Jular. (2020). *Treehouse modular houses*. Retrieved from https://www.jular.pt/en/produtos/casas-pre-fabricadas/casas-modulares-treehouse.
33. MIMA. (2020). *MIMA housing*. Retrieved from https://www.mimahousing.com/.
34. MaisrArquitetos. (2019). EVOA—Environmental Interpretation Center. *ArchDaily*. ISSN 0719-8884. Retrieved from https://www.archdaily.com/316836/evoa-environmental-interpretation-center-maisr-arquitetos.
35. Modular System. (2020). Ecocamp Cocoon. *Comporta*. Retrieved from http://www.modular-system.com/en/projects/ecocamp-cocoon/.
36. Jular. (2020). *Zmar Eco Campo Resort*. Retrieved from https://www.jular.pt/referencias/turismo/zmar-eco-campo-resort.
37. de Andrade, L. R., & Aguiar, D. (2012). Pedras Salgadas Eco-Resort. *ArchDaily*. ISSN 0719-8884. Retrieved from https://www.archdaily.com/307297/pedras-salgadas-eco-resort-luis-rebelo-de-andrade-diogo-aguiar.
38. Modular System. (2020). *Beach Villas—Pestana Tróia Eco Resort & Residences*. Retrieved from http://www.modular-system.com/en/projects/beach-villas/.
39. Tavares Pereira, L. (2020). Mountain House, Louredo, Vieira do Minho - [A] ainda arquitectura. Text provided by the authors.
40. Portugal. (2014). *Plataforma de Acompanhamento das Relações nas Fileiras Florestais* (Despacho no. 8029/2014).
41. PNAM. (2020). *Prémio Nacional de Arquitectura em Madeira*. Retrieved from https://pnam.pt/.
42. Pereira, M. (2019). Conheça as obras finalistas do PNAM'19—Prêmio Nacional de Arquitetura em Madeira. *ArchDaily Brasil*. ISSN 0719-8906. Retrieved from https://www.archdaily.com.br/br/927703/conheca-as-obras-finalistas-do-pnam19-premio-nacional-de-arquitetura-em-madeira.
43. Menos é Mais. (2020). Descriptive memory—To build without touching. Text provided by the authors.
44. do Vale, T. (2020). The rejection of wood as a building material. Text provided by the author.
45. MaisrArquitetos. (2020). Why architects and builders are not using wood in their buildings?. Text provided by the author.
46. Appleton, J. G. (2005). *Reabilitação de edifícios gaioleiros*. Amadora: Edições Orion.
47. Tavares Pereira, L. (2020). Why architects and builders are not using wood in their buildings?—[A] ainda arquitectura. Text provided by the authors.
48. Regino Cruz. (2020). Why are architects and builders not using wood in their buildings? Text provided by the author.

Chapter 7
Sustainable Wooden Construction of Traditional Houses in Moderate Humid Climate of North Iran

Seyedehmamak Salavatian

7.1 Introduction

Rural architecture is the product of three counter-factors: environmental ecology, social ecology, and individual ecology. Dependence on the environment is a common feature in vernacular architecture which is observed in both the appearance and logic of architecture. Materials are one of the essentials in this regard. Wood is the oldest material after stone used by humans for construction. It is a renewable and naturally grown building material with a rich history while providing valuable opportunities for the building industry of the future [1]. Wood has been a prevalent material before the industrial revolution and about 80% of buildings were made of wood. By the appearance of synthetic materials as iron, steel, and concrete, the traditional place of wood in construction became questioned [1]. The predominance of steel and reinforced concrete led to a gap in the application of wood in research and practice.

However, the climate change issue and environmental concerns in the recent decays necessitated major changes in conventional construction systems and laid the groundwork for increased use of renewable materials like wood. Besides, the newly adopted building regulations were notably caused the growth in the construction of multistory wooden buildings.

For centuries, the use of wood was solely limited to simple rectangular members, while in the last century, better use and management of wood became possible due to the advances in wood production technology as decomposing and reconstructing at a cellular level. Thus, its application in buildings transformed from linear

S. Salavatian (✉)
Department of Architecture, Rasht Branch, Islamic Azad University, Rasht, Iran
e-mail: salavatian@iaurasht.ac.ir

A. Sayigh (ed.), *The Importance of Wood and Timber in Sustainable Buildings*, Innovative Renewable Energy, https://doi.org/10.1007/978-3-030-71700-1_7

members to planar ones. Using fiberboard, plywood, and cross-laminated timber provided new opportunities for its utilization in modern architecture.

Besides, the environmental superiority of wood over other prevalent materials is demonstrated in previous studies. According to Gordon, the production of a panel with a given compressive strength requires 500 times less energy than steel. Wood also offers remarkable physical and structural benefits. Even though wood has almost the same compressive strength as concrete, it has a significantly smaller heat transfer coefficient. Wood is also lighter than steel for the same tensile loading capacity [1].

Today, about 30% of all land area worldwide is covered with forests about 57% of which belong to developing countries [1]; therefore, there are worthy potentials to flourish the effective usage of wood in the building sector in such countries. However, investigation in environment-friendly materials including the wood industry should be encouraged. Wood construction is not restricted to rustic buildings, and its merits for high-technology buildings should be introduced to developers and clients. In addition, wood's natural beauty and warmth have shown positive effects on occupants which is a principal issue in modern buildings nowadays. As a visual effect, wood lowers the sympathetic nervous system which is responsible for psychological stress responses in humans.

Wood has been widely used in traditional houses of North Iran and is a source of inspiration for sustainable building practice. In this study, its application in the aforementioned buildings is investigated, and new perspectives are illustrated for its usage in future sustainable buildings of similar climates.

7.2 The Role of Wood Material in the Studied Geography

Country of Iran involves four main types of climate conditions. The studied subject in this chapter is the wooden architecture of Guilan province (about 9% of the whole country area) located in the southern side of the Caspian Sea and northern side of the Alborz Mountain with a moderate and humid climate. As seen in Fig. 7.1, the entire province is divided into distinguished regions, including mountainous, foothill, plain, and coastal [2]. The determining features of this climate are the extensive coverage of vegetation, intense precipitation, and high percentage of humidity throughout the year [3].

Materials in village houses are mainly vegetal and soil-based with minimum damages to the environment. No garbage remains in this system, and materials change from a useful state to the other. Lateral products are applied for different services and ultimately return to the earth.

Despite wood, materials like clay, as the filler for spaces among branches; stone, to level the ground surface; and straw, as an addition to mortar for better adhesiveness, are common as well. Also, wood by-products—coal, ash, etc.—are utilized in foundations; they are mixed with soil and rammed [4].

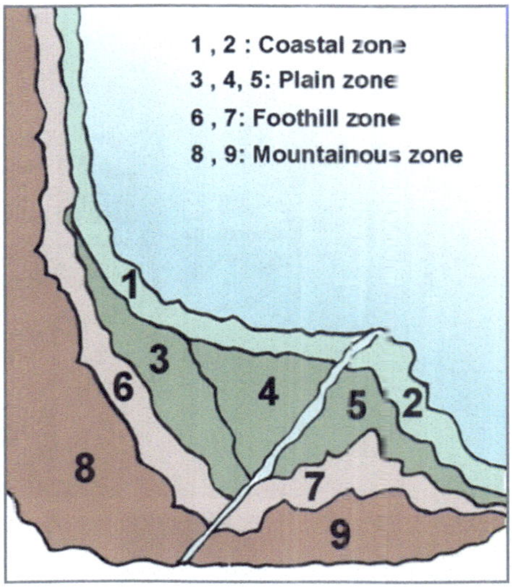

Fig. 7.1 Various geographic divisions of Guilan province

Various regional trees with the appropriate moisture resistance are employed. Woods made of Mountain cedar trees are used as building elements due to their abundance in the region. Particularly, in foothill zones, all the principal building components including floors, skeleton, and roofs are built of wood which provide the required tensile and compressive strength.

The wooden members are applied in the ax cutting forms. Wood is cut during the autumn since it is solider than in spring which provides more resistant material. In order to dry wood in traditional method, it is stored in a covered place far from rain and moisture for about a year; if the drying process does not complete, it extends to maximum 2–4 years depending on the wood condition [5].

Wood also has some weak points which must be considered; it is exposed to decay caused by fungi, insects, and moisture. Humidity as the most important factor could make continuous expansions and contractions which result in problems as distortions in different directions (Fig. 7.2).

7.3 Traditional House Types

Rural plans in most villages of this area are scattered; in other words, buildings are placed distant from each other to facilitate air movement among the masses. Architectural configurations and spaces layout in the areas shown in Fig. 7.1 differ from each other, and this difference stems from the variances in climate, topography, and people's livelihood. Their typological analysis and sustainability principles have been previously studied by author [6]. Due to the richer architectural features of traditional village houses in plain and foothill areas, wooden architecture in these two zones is the subject of study in this chapter. Examples of houses in each zone are illustrated in Fig. 7.3.

Fig. 7.2 Examples of using vegetal and by-product materials for various purposes (Guilan Rural Heritage Museum, Saravan, Guilan)

Fig. 7.3 Examples of traditional houses. (**a**) Mousazadeh House (plain zone), (**b**) Jame House (foothill zone) (Guilan Rural Heritage Museum, Saravan, Guilan)

Guilan Rural Heritage Museum located in the forest park of Saravan is an open-air museum established to preserve the architectural heritage of this region. The valuable houses found in all the zones presented in Fig. 7.1 were dismantled and reconstructed in the eco-museum [7]. These preserved houses are the scope of study in this chapter.

Plain Area

Buildings in plain areas are usually constructed in two-story levels and elevated from ground by a crawl space [8]. The overall form of houses is simple, pure, and longitudinal. Rooms are based on a square-shaped module and are set in a linear geometry. The buildings façades are quite transparent due to the higher level of humidity and necessities for natural ventilation. These transparent enclosures surrounding the building core provides spaces turning around or at least two sides of the building as shown in Fig. 7.4. Thus, porches and terraces are key elements of houses carrying a lot of activities. They are used about 9 months a year, and their width is about 2–2.5 m, providing an effective linkage among rooms and also connecting different floors in multilevel houses. Southern porches are larger than other sides comprising more daily activities with greater wooden ornaments [9]. An example of spacious terrace is illustrated in Fig. 7.5.

Foothill Area

In this area, the family economy is based on livestock and agriculture. The ground level in houses is normally allocated to cattle and poultry. This setting assists in keeping the upper floor warmer in cold seasons. In foothill areas, the relative humidity is lower, and the weather is colder compared to the plain zones. Therefore the level of transparency lowers; semi-open spaces on the first floor are limited solely to one façade.

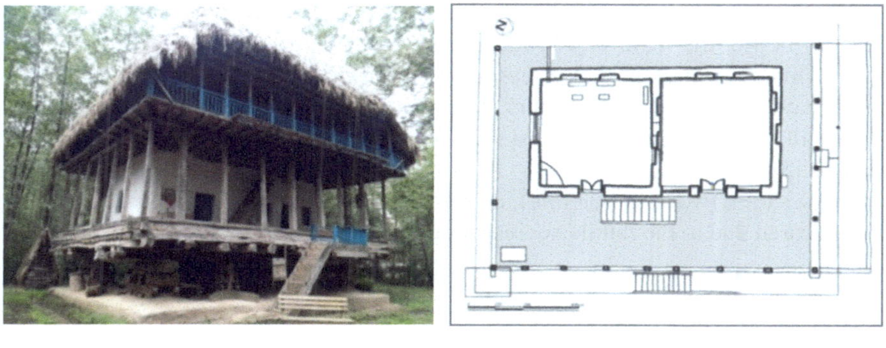

Fig. 7.4 Rafiyi House, plain zone [4]

Fig. 7.5 Southern terrace, Mousazadeh House (Guilan Rural Heritage Museum, Saravan, Guilan)

Fig. 7.6 Behzadi House, foothill zone [5]

Plan proportions vary compared to the plain areas; plan geometries are square-shaped rather than rectangular ones in plain zones. This difference comes from the greater need for heating energy and preservation of heat inside.

The main axis in houses is turned northeast/southwest. For example, in the Behzadi house, the main façade faces the east with a rotation of 10° toward the north. This orientation provides seasonal wind from the northeast in summers and solar heat from the south in winters. The west side is devoid of any openings, and thick walls (up to 50 cm on the west side) prevent heat loss and severe winds in cold seasons [5].

Houses have two stories at most, and two-, three-, four-, and five-bedroom types are appeared due to the family socioeconomic status (Fig. 7.6).

7.4 Wooden Building Members

Different wood types and products are applied for various purposes. Wood derived from *Aluns glutinosa* trees is used for members protected from moisture, such as floors or bearing components and trusses [4]. *Samad* is also a common tree type of the region suitable for foundations, columns, and stairs due to its high strength. *Zelkova* is another wood type much in demand with an acceptable resistance [10]. *Tabrizi* is also a proper wood type for ornaments and balusters [2].

No nails are utilized for wooden pieces of walls and handmade nails used for columns. Tongue and groove joints are used for most conjunctions: wooden piece surfaces are half cut, and the cut ends are put together. Where the wooden member is not long enough, other pieces are used as extensions; in these cases, the heads are beveled by the angel of 45° for better interconnection [5]. In the following sections, the application of wood material in four main building elements as structural components, walls, roofs, and ornaments is discussed.

Structure

Wooden Foundations

Wooden foundations are among the noteworthy construction systems in local houses of this region which have shown acceptable performance. The earth type is composed of weak soil, and also the high groundwater level might cause problems; therefore, to prepare a proper bed, the earth dug to the depth of 1–1.5 m. Then, the ditch is filled with gravel or rubble stone and rammed layer by layer to provide a firm bed.

First, a log made of mulberry tree with the estimated diameter and length of 35 and 100 cm, respectively, is used along the longitudinal axe of the building and set over the bed. Upon the first layer and in the perpendicular direction, three logs are put with a diameter of 25 cm made of available local and resistant wood. For the next layer, two pieces of lumber (diameter of 30–35 cm) perpendicular to the layer beneath and parallel to the lowest one are set. As the top layer of the foundation, a thick timber with a trapezoidal section is used and bears the main building load directly (Figs. 7.7 and 7.8). All the details described above are exposed, and the structural behavior of the building is intelligible

There is no solid linkage among the abovementioned components, and the form of the truncated pyramid of the foundation bears the compressive load. The lack of any rigid connections causes subtle movements of the building against lateral forces of earthquakes providing the overall stability of the building. Two rows each including five wooden foundations are sufficiently used for medium-sized two-story houses in the plain zones [11].

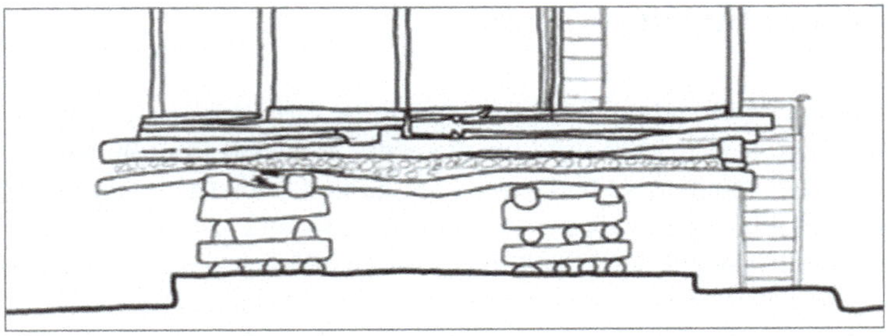

Fig. 7.7 *Shekili* foundation detail [2]

Fig. 7.8 Construction of *shekili* foundation, Por Mehr House (Guilan Rural Heritage Museum, Saravan, Guilan)

Wooden Skeleton

The load-bearing system in vernacular houses in this area, including both plain and foothill zones, is based on the wooden members. As the wall arrangement reaches the considered height for the ground floor ceiling, beams are set; wooden beams with the estimated length of 5 m and in the distance of 60 cm from each other are used along the room width and rest over the opposite walls. This way, the ceiling load is transmitted to the main girder—*"Naal"*—and the bearing walls. Finally, boarding over the beams is applied, and the final finishing by clay mortar is carried out (Fig. 7.9).

Projection of secondary beams of ground floor ceiling—*"Vashan"*—provides an edge in the upper floor; besides the front row of posts, there is another row of vertical wooden posts along the columns in the ground floor and forms a corridor between two rows of wooden posts made of Zelkova trees at the height of 2.20 m.

Again, as the arrangement of timbers to the final height of the first level completes, the main girder—*"Naal"*—is installed over the columns, and secondary beams—*"Vashan"*—of the first level ceiling are rested at any 1 m over the room walls and the *"Naal"* as shown in Fig. 7.10.

Fig. .7.9 Main structure members, Jame House (Guilan Rural Heritage Museum, Saravan, Guilan)

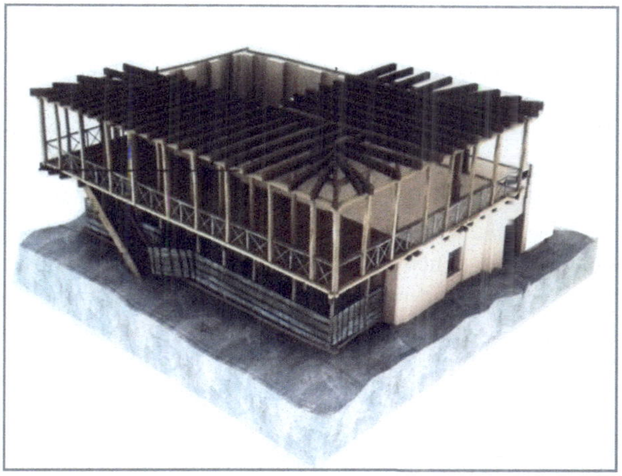

Fig. 7.10 The overall structural schema [10]

The sloped roof load is also transmitted from diagonal elements to the secondary beam—"*Vashaan*"; then, it is carried by the main girder over the columns, "*Naal*," as well as the bearing walls. Finally, the load conveys to the foundation by the ground floor columns.

Some other wooden rods are utilized for miscellaneous purposes; in terraces, wooden rods at the level of 170–180 cm—called *Dastak*—suited for hanging local crops. Also, two parallel wooden rods with an estimated distance of 10 cm at the height of 170–180—called *Dudan Dar*—are built for hanging and drying wheat plants.

Walls

The wall construction methods in practice for exterior walls and major interior walls are described as below:

Orjon: in this construction method, slates are set and leveled at the wall positions; then, timbers are rested over the slates, and their end points are carved at the conjunction to the adjacent side to provide a proper leaning surface (Fig. 7.11). For longer walls, two 4-m timbers are put along each other. In order to connect the extended timbers, two shorter pieces (1 m each) are applied in a perpendicular direction. These short ones are held by other pieces (1.2 m each) parallel to the main wall. This anchorage as shown in Fig. 7.12 continues for the whole height of the ground floor. The wooden exterior walls of the first floor are also installed along the walls of the lower level.

The spaces between wooden members are filled with clay. This mortar layer protects wood from insects and fungi and also provides a relative connection among wooden members. The mortar thickness and timber arrangement are in such a way that the wall shrinks vertically inward; in other words, the room floor area is larger than the ceiling area.

In the houses of the plain zone, after the construction of wooden walls in the abovementioned method, wall surfaces might be covered by the clay mortar mixed with "*Kolush*"—rice stalk—and "*Fal*"—rice flake—as the mortar adhesive. The finishing process replicates after a few months to totally cover the seams and cracks.

Darvarchin: Trunks in circular sections are stacked every other one, and joints are made in a concave shape to connect each other strictly [5]. A local plaster made of clay, water, and rice flake might be applied to the final surfaces. This method is mostly used for one-room buildings; the underneath ground is dug and filled with cobblestone to protect the wood from humidity.

Fig. 7.11 An example of *Orjon* wall type, Behzadi House (Guilan Rural Heritage Museum, Saravan, Guilan)

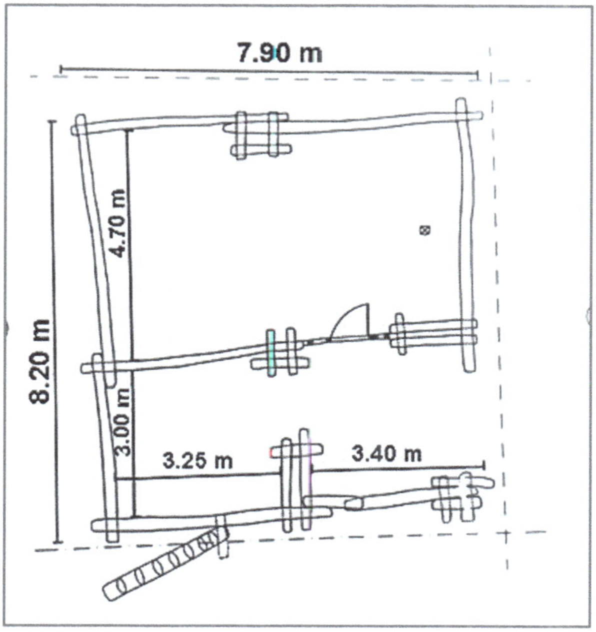

Fig. 7.12 Anchorages of wooden bearing walls [5]

Fig. 7.13 Wall types; (**a**) *"Darvarchin,"* (**b**) *"Nefar"* (Guilan Rural Heritage Museum, Saravan, Guilan)

Nefar/Zegali: in this construction system, two sides of pillars are enclosed by thin branches in the horizontal or diagonal direction, and the space between is filled by clay [5] (Fig. 7.13).

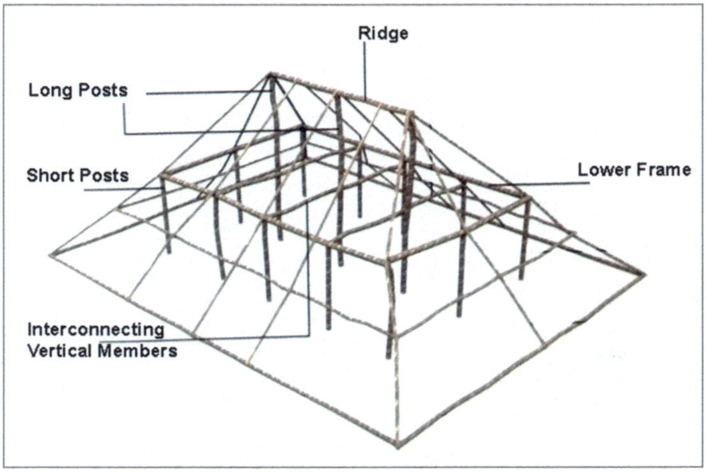

Fig. 7.14 Roof structural elements [10]

Roofs

Due to the considerable precipitation in this region, hipped roofs are the conventional roof type for houses in this region. The overall roof structure consists of a frame in the lower height (about 2–2.20 m) and a row of longer posts in the middle line as shown in Fig. 7.14. The shorter frame consists of smaller posts installed at the distance of 2 m inward the building perimeter. Above both the shorter and longer posts, horizontal members are put. Moreover, other horizontal members perpendicular to the ridge link the facing short posts [10].

Roof covering is mostly made of wood in foothill zones, called *Lat* (Fig. 7.15)*;* pitched roofs are covered by small wood pieces with the estimated size of 40 × 20 cm made oak slats placed above each other as each piece covers 5 cm of the pieces underneath [5]. However, rice stalk and vegetal fibers are more frequent in plain areas.

Ornaments

The tendency to the beauty is an innate quality, and human being has always tried to create and promote beauty in his artifacts. Decorative architectural elements make buildings more pleasant as well. Decorations not solely are a superficial covering but also include symbolic levels which link to the cultural, national, and geographical features of each region. Building ornaments vary depending on the owners' social and economic status. Residents of traditional houses particularly in rural areas spend much time outside and in open spaces during the day; thus, buildings are less decorated than houses in the country's central plateau in general.

Fig. 7.15 Wooden roof covering *"Lat"* (Guilan Rural Heritage Museum, Saravan, Guilan)

In this area, ornaments are mostly based on wood material. Capitals, beams, and balusters are the subject of wooden decorations which among them the balusters are the key elements applied in both first and ground levels on one or all sides Terraces and semi-open spaces are of great importance in villagers' lifestyle and in some cases are the unique decorated spaces of the house.

Surrounding nature has been the source of inspiration for ornament patterns. Geometric, vegetal, animal, and celestial patterns as well as inspirations from daily objects have been used widely [12] as shown in Table 7.1. In west side villages, terraces parapet decorations are denser containing less voids in order to provide better climatic protection as well as privacy.

7.5 Conclusion

Studies have shown that the traditional architecture of Guilan province is in line with sustainability principles. Due to the climatic features and the abundance of local wood and vegetal materials, buildings are mostly made of wood, which was reviewed in this chapter.

In the overall structure, column-beam systems as well as wooden bearing walls—containing clay fillers—have the mission to transfer loads to the totally wood-made foundations and finally the ground. In two-story wooden buildings, terrace wooden columns bear the weight of upper terraces.

Wooden horizontal beams in flat middle ceilings and vertical or diagonal beams in the final sloped roof convey loads to the girders and bearing walls. Floor pavings are fabricated by wooden boards, and final roof coverings in many foothill cases are built of wooden pieces.

Table 7.1 Wooden ornaments of terrace parapets (Guilan Rural Heritage Museum, Saravan, Guilan)

No	Name	Description	Image
1	*MohtashemTalab* House	**Geometric patterns** Such as diamond, square, and zigzag shapes	
2	*Rostami* House West Plain zone	**Vegetal patterns** Inspired by pine tree, cedar tree as the symbol of eternity, and motifs resembling pomegranate branches as the symbol of fertility	
3	*Sadeghi* House Central Plain zone	**Animal patterns** Butterfly/cross-shapes, fish/fishbone shapes	
4	*Hasani* House West Plain zone	**Celestial patterns** Such as the sun rays and star design	
5	*Amini* House Central Plain zone	**Daily object pattern** Such as flower vase and hair comb	

In this chapter, the fully wooden architecture of the region was investigated in order to provide new standpoints for wood application combined with today's advanced technology.

References

1. Menges, A., Schwinn, T., & Krieg, O. D. (2017). *Advancing wood architecture*. New York: Routledge.
2. Taleghani, M., & Amini House. (2010). *Guilan rural architectural heritage*. Tehran: Rozaneh.
3. Ghobadian, V. (1995). *Climatic analysis of the traditional Iranian buildings*. Tehran: Tehran University.
4. Taleghani, M., & Rafiyi House. (2007). *Guilan rural architecture heritage*. Tehran: Moein Publisher.
5. Taleghani, M., & Behzadi House. (2011). *Guilan rural architecture heritage*. Tehran: Rozaneh.
6. Salavatian, S., & Asadi Malekjahan, F. (2019). Typological analysis of vernacular residential buildings in moderate-humid climate of North Iran. In *Sustainable vernacular architecture* (pp. 115–140). Cham: Springer.
7. *Guilan Rural Heritage Museum* [Online]. Retrieved September 1, 2020, from http://gecomuseum.com/fa/Default.aspx.
8. Khoshsima, E., Mahdavi, A., Inangda, N., & Rao, S. P. (2009). Evaluation of traditional architecture of southern shores of Caspian Sea Region in Iran. In *Conference of the International Journal of Arts and Sciences*, 2009.
9. Memarian, G. (1992). *An introduction to Iranian residential buildings (extrovert typologies)*. Tehran: Elm o Sanat University.
10. Taleghani, M., & Mousavi House. (2010). *Guilan rural architecture heritage*. Tehran: Matn Artwaork.
11. Khakpour, M. (2006). Construction of Shekili Houses in Guilan. *Honar-Haye-Ziba, 25*, 45–54.
12. Alizadeh, S., & soheili, J. (2017). Typology of ancient architectural designs in Gilan in accordance with the role of some handmade arts of this region with emphasis on Peirce ideal. *Islamic Art Studies, 13*(26), 75–104.

Chapter 8
Multiple Scales Insight into Using Timber for a Sustainable and Future Approach to Buildings

Carolina Ganem Karlen

8.1 Trees as a Sustainable Resource

Wood is the only significant building material that is grown. Owing to its unique characteristics, wood has historically been a valuable and useful natural resource. It is also one of the most important construction materials mankind has ever come across.

The most abundant tree-forming groups are the angiosperms (a group of plants producing flowers and enclosed seeds: dicotyledonous, often deciduous broad-leaved, including oak, birch, beech, and ash) and the gymnosperms (plants producing uncovered seeds such as spruce, pine, and fir). Industrially, wood obtained from angiosperms is called hardwood and that from gymnosperms softwood. This nomenclature does not necessarily reflect the actual wood properties; balsa (a hardwood) is much softer than average softwood.

Trees of the same species grown in diverse conditions may grow very differently. The increase in girth of a tree grown in the open is twice as much as one grown in woodland. An average free-standing tree would add 2.5 cm per year to its girth, with fast trees (like giant sequoia, coastal redwood, Sitka spruce, and Douglas fir) reaching 5–7.5 cm. Some trees, for example, Scots pine, grow more slowly [1].

The growth of a trunk is achieved by two kinds of events, each controlled by specialized parts of the plant. The first is mediated by the shoot (*apical meristem*) located on the top of the tree and is responsible for predominantly upward primary growth. This growth is common for all vascular plants. Trees also have the ability of secondary growth (In contrast to non-tree plants), which means that their stems can get thicker. This growth is determined by the proliferative activity of vascular

C. Ganem Karlen (✉)
INAHE - CONICET and FAD - UNCUYO, Mendoza, Argentina
e-mail: cganem@mendoza-conicet.gov.ar; carolinaganem@gm.fad.uncu.edu.ar

A. Sayigh (ed.), *The Importance of Wood and Timber in Sustainable Buildings*, Innovative Renewable Energy, https://doi.org/10.1007/978-3-030-71700-1_8

cambium, a group of dividing cells located between and giving rise to the xylem (water-conducting tissue positioned on the inside of the trunk) and to the phloem (tissue responsible for the transfer of nutrients and situated on the outside of the trunk). The molecular mechanisms regulating wood formation are the subject of intensive research [2]. Matching optimized growth and the usefulness of trees for construction is not an easy task. Importantly for construction, heartwood and sapwood have different properties. Wood is nonuniform within heartwood and sapwood layers. See Fig. 8.1a, b.

In general, trees produce annual rings. Such rings reflect the changing environment. Rapid growth during the spring produces early wood which is less dense and composed of large cells with thinner walls allowing for efficient water transport to support intense photosynthesis. This period is followed by slower growth, yielding

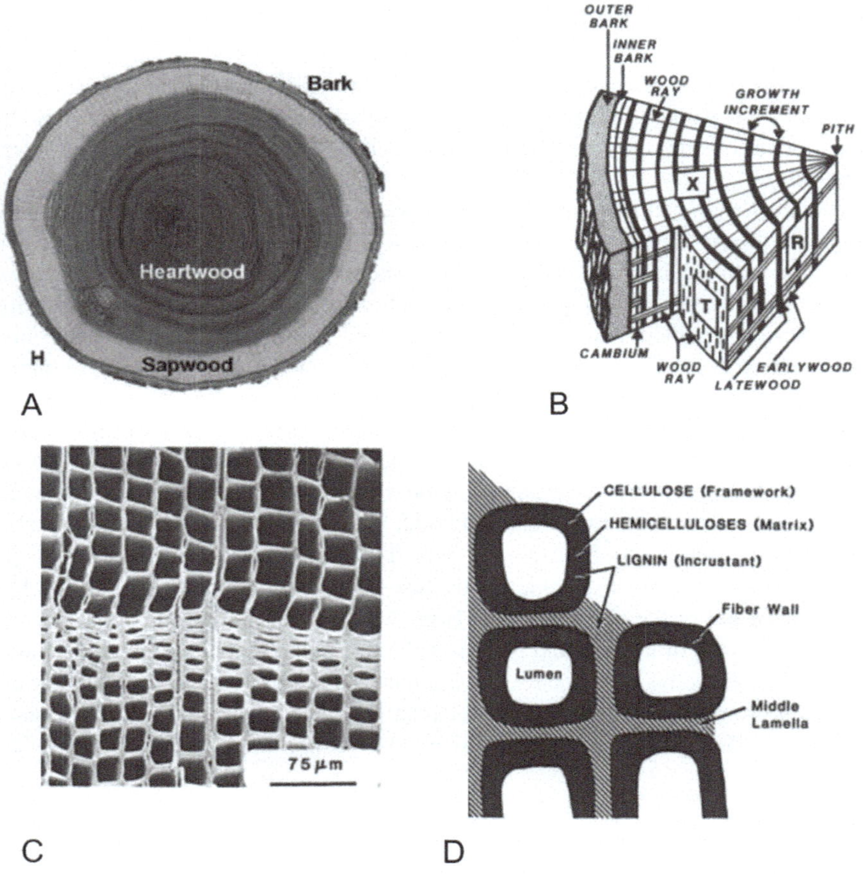

Fig. 8.1 (**a**, **b**) Tree cross section cut to reveal the three major structural planes of wood bark, sapwood, and heartwood. (**c**, **d**) Cross-sectional view of fully differentiated wood cells and fibers [3]

latewood, characterized by more densely packed, smaller cells, production of which stops for winter. Each annual ring consists of both early and latewood [4]. Wood contains carbon, which is part of the glucose that constitutes cellulose and part of the hemicellulose of the cell wall. See Fig. 8.1c, d. The attachment of carbon dioxide (CO_2) in wood refers to the amount of CO_2 that the tree needed to take from the atmosphere to fix the carbon through photosynthesis in the cell wall and release the oxygen back into the air.

Now, the cell wall of wood has not only glucose (container of fixed carbon) but also other substances, which is why 1 kg of wood will contain considerably less than 1 kg of carbon: for every kg of carbon contained in the cell wall of the wood, the tree requires fixing 3667 kg of atmospheric CO_2, but each kg of wood contains in its cell walls approximately 0.444 kg of Carbon; therefore, to form 1 kg of wood, the tree requires taking from the atmosphere approximately 1.63 kg of CO_2 (1.63/0.444 = 3.667) [5].

In the case of referring to the CO_2 content per volume of wood, the density of the wood is very important. The higher the density, the higher the carbon content and consequently the CO_2 set. According to the IPCC, the average carbon content of conifers expressed in kg is 50% of the biomass. Biomass (kg) is in turn the product of the volume of natural wood by the density of it [6, 7].

8.2 Timber Properties as Construction Materials

Wood and other cellulosic materials are poor conductors of heat because wood and its derived products are porous bodies, and therefore, their conductive amplitude has an intermediate value between those of its solid components and those of the air contained in pores. It is necessary to indicate the direction of propagation of the heat flux. The reason for this difference we must look for it in the microstructure of wood, since the transport of heat due to conduction encounters lower resistance in the direction of long cellulose chain molecules, which are orthogonal to them [8].

Thermal conductivity declines as the density of the wood decreases. In the direction of the grain, the thermal conductivity of wood is about twice what it is perpendicular to the grain. For example, the thermal conductivity of pine in the direction of the grain is 0.22 W/m °C and perpendicular to the grain 0.14 W/m °C.

Increasing the moisture in the wood also increases its thermal conductivity. As the temperature of wood decreases, its strength usually increases. The thermal expansion of wood in the direction of the grain is very little. In the radial and tangential directions, temperature movements are much greater. The relationship between the thermal expansion coefficients and moisture contraction coefficients of wood in different directions relative to the grain is in the same class in terms of size. Repeated variation in temperature decreases the strength of wood. At a temperature of less than 0 °C, wood may start to crack as water in the cell lumens expands as it freezes.

Fig. 8.2 (**a**) First column, from top to bottom: examples of softwood and hardwood leaves and seeds. (**b**, **c**) Light microscopic views of the lumina (L) and cell walls (arrowheads) of a softwood (**b**) and a hardwood (**c**). (**d**, **e**) Hand-lens views of growth rings, each composed of early wood (ew) and latewood (lw), in a softwood (**d**) and a hardwood (**e**) [8]

Also, there are significant cellular differences between angiosperms and gymnosperms. In softwoods, tracheids are the predominant wood cell. In hardwood, there are two primary types of wood cells: fibers (constituting 50% of the wood volume) providing structural support and water-conducting vessels (30% of wood). See Fig. 8.2.

The layered structure of the wood cell wall is a major determinant of strength and mechanical properties. Wood is highly anisotropic, meaning that its physical properties differ along different axes. The angle between the cellulose microfibrils and the longitudinal cell axis, the microfibril angle, is found to be a critical factor in determining the structural and mechanical properties.

The structure of a wood cell wall is that of a multilayered composite as shown in Fig. 8.3. The xylem tracheid (softwood) or fiber (hardwood) cell wall has four distinct cell wall layers (primary, S1, S2, and S3).

Between two adjacent cells lies a highly lignified region called the middle lamella. The middle lamella, a lignin-pectin complex, is responsible for cementing the cell walls of two adjoining cells together. The primary cell wall is a thin layer, and the microfibrils are deposited in a random fashion. Both the middle lamella and adjoining primary walls are sometimes referred to as the primary layer [9]. In the secondary cell wall layers, the microfibrils are closely packed and parallel to each

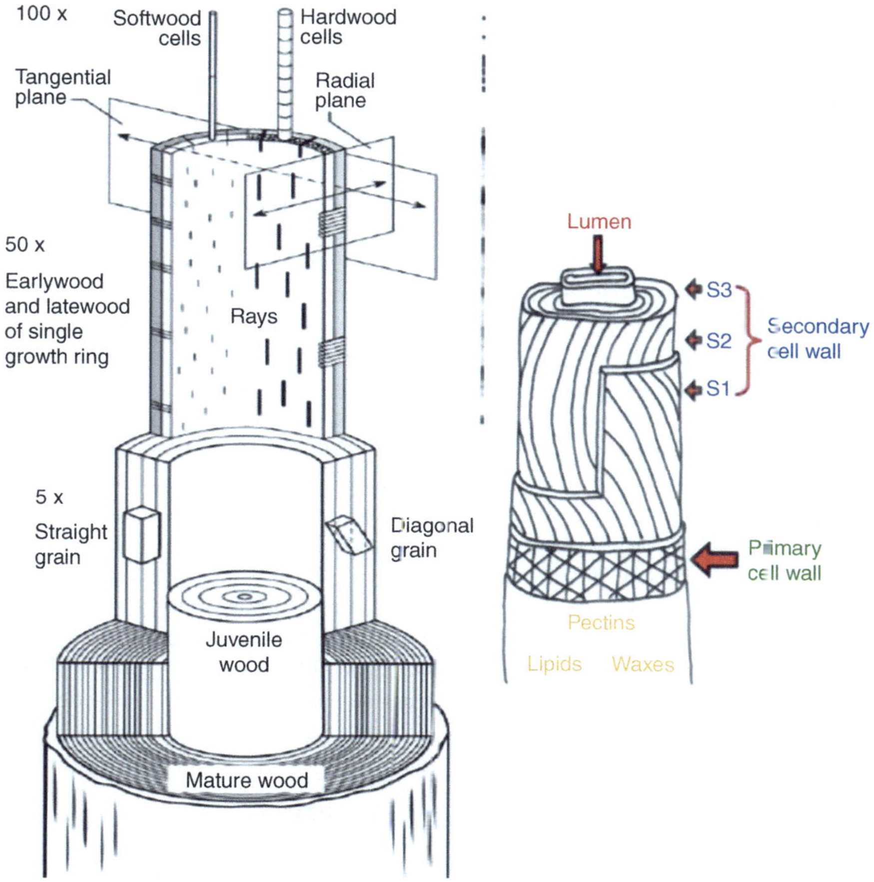

Fig. 8.3 Cell wall structural organizations at various magnifications. (**a**) [8] and (**b**) [4]

other. In addition to the cell lumen, the secondary cell wall is subdivided into three layers: S1, S2, and S3 [10].

Fig. 8.3a illustrates a cut-away tree at various magnifications. At the top, at an approximate magnification of 100×, a softwood cell and several hardwood cells are illustrated, to give a sense of scale between the two; one tier lower, at an approximate magnification of 50×, is a single growth ring of a softwood (left) and a hardwood (right), and an indication of the radial and tangential planes; the next tier, at approximately 5× magnification, illustrates many growth rings together and how one might produce a straight-grained rather than a diagonal-grained board; the lowest tier includes an illustration of the relative position of juvenile and mature wood in the tree, at 1× magnification [8].

The varying fibril orientation in the particular layers (50–70° in S1, 5–15° in S2, and 60–90° in S3) causes a mechanical locking effect, leading to very high stiffness of the overall cell [11]. Due to its thickness (90% of the secondary cell wall) and low value of microfibril angle, the S2 layer is responsible for the high tensile strength and stiffness and low shrinkage of wood in the longitudinal direction. Increased microfibril angle in the S2 layer decreases cell wall tensile strength and stiffness but increases durability. This enables trees to adjust both stiffness and toughness of the tissue by shifting the cellulose fibril orientation of the cell wall. The S1 and S3 layers are thin, and the microfibrils of each alternate between left- and right-hand spirals. The fibril angle and thickness of the S1 and S3 layers are believed to be of significance to mechanical properties in the transverse direction [12].

Timber is one of three structural materials currently used in the construction of large structures, along with steel and reinforced concrete. If timber is used in the types of building in which it is most structurally efficient, then the timber we harvest can do the most to reduce the environmental impact of construction. Timber has strength parallel to grain similar to that of reinforced concrete: hardwood is slightly stronger and softwood slightly weaker, although timber cannot match modern high-strength concrete in compression. Timber is less stiff than concrete, and both materials are far less stiff and strong than steel. However, timber has a low density compared with these other conventional structural materials. This results in efficiency for long-span or tall structures, in which a significant part of the load to be carried by the structure is its own weight [3].

When those loads are resisted purely in tension or compression, the strength-to-weight or elastic modulus-to-weight ratios are measures of the mass of material required to achieve a structure of a given area, height, or span. Figure 8.4 shows compression strength and modulus of construction materials normalized by density,

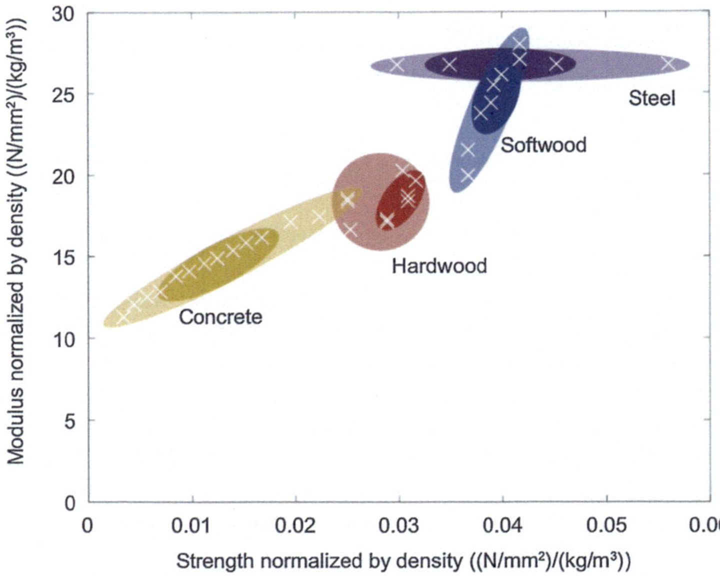

Fig. 8.4 The strength-to-weight and modulus-to-weight ratios for steel, timber, and reinforced concrete shows that softwood performs similarly to steel by those measures [3]

according to the relevant design standard. The dark colors represent the more widely used grades of material. Design values of strength and stiffness, including partial factors, are shown based on the Eurocode design standards for concrete (BSI. BS EN 1992-1-1:2004) [13], steel (BSI. BS EN 1993-1-1:2005) [14], and timber (BSI. BS EN 338:2009) [15].

This suggests that timber is a particularly structurally efficient material in structures or parts of structures, in which a high proportion of the load to be resisted is the self-weight of the structure itself. It has also good seismic performance due to its lightweight, and even if timber elements are not able to have a ductile behavior, using steel connection allows building dissipative structure.

8.3 Cultural Practices to Use of Timber for Building in Different Time Periods

Wood is one of the longest-standing building materials in existence, with evidence showing homes built over 10,000 years ago used timber as a primary source for construction materials.

The first timber homes date back to the Mesolithic period and were found at Killerby Quarry, Britain, preserved in peat. One was found in a layer dating to the later Mesolithic or Neolithic period (6000–3000 BC), while the other, identified in a lower, early Mesolithic level (dating to c. 10,000 BC). Thanks to the anaerobic conditions of the peat in which the timbers had lain for thousands of years, they were so well preserved that the marks of the stone axes used to prepare and trim them could still be seen. See Fig. 8.5a.

Fig. 8.5 From left to right. (**a**) Remains of Mesolithic timber structures, thought to be the remnants of teepee-like huts [16]. (**b**) A 10,000-year-old Mesolithic hut recreated by University College, Dublin [17]

The younger of the two structures (in the upper layer) comprised two Y-shaped timber poles, about 6 m long, used to construct an A-frame shape. It is thought that they could be part of a temporary encampment that was built on the edge of a pond and used during the summer. Of the older structure, five 6 m- to 7 m-long poles have survived, although the original total is unknown. They were situated around the edge of a roughly circular hollow, in a way that suggests that they may have produced a cone-shaped roof, which could have been covered with animal hides, reeds, or other materials. The well-preserved remains of a fire were also found immediately below the collapsed roofing timbers within the hollow [18]. Figure 8.5b shows a recreation of a 10,000-year-old Mesolithic hut.

From 9000 BC to 5000 BC, one of the largest structures in the world was the Neolithic longhouse, a long narrow timber structure housing 20–30 people. In the wider area of Central Europe, Neolithic longhouses have been associated with the first agricultural populations since the time of their inception.

The sedentary nature of life attached to a specific area represents a crucial civilizational change. Settling in one place enabled people to build more permanent and stable homes, which can also be considered as being houses in the modern sense. They were therefore spaces that provided not only a certain amount of comfort and a suitable background for both the material and spiritual aspects of human existence but also a place for production or for storage during unfavorable weather seasons.

The remains of these prehistoric houses are recorded archaeologically, only in the regard to the recessed portions of their ground plans. These in fact represent the negative imprints in the subsoil of base postholes or troughs, in which originally wooden poles were erected. See Fig. 8.6.

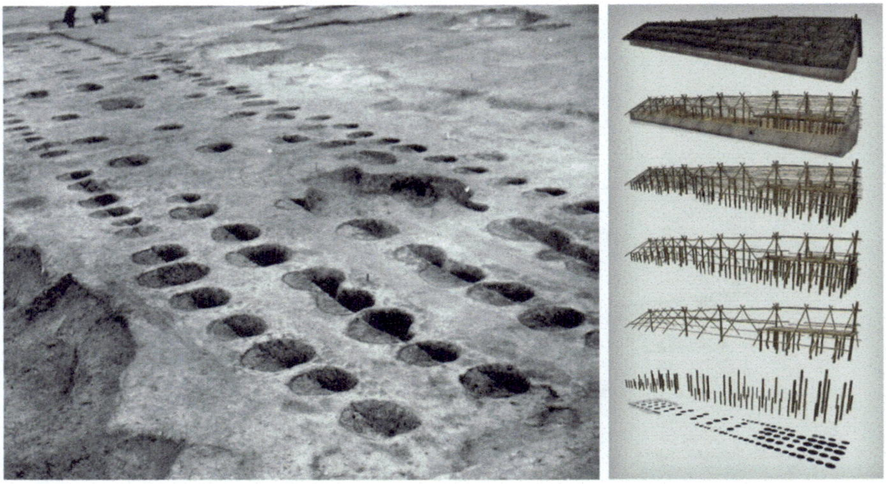

Fig. 8.6 From left to right. (**a**) Photograph of the remains of the ground plan of a Neolithic longhouse located at the Bylany site. (**b**) Hypothetical reconstruction of a longhouse of the Linear Pottery culture based on plan No. 41 from Bylany, Czech Republic [19]

The oldest houses of the Linear Pottery culture accounted for the aboveground structures that consisted of five rows of load-bearing columns. The outer walls of the house were interlaced with wicker and subsequently daubed, using a mixture of clay and such organic ingredients as grass and straw. The center row and the two inner rows of posts propped up the roof truss and the purlins. The timber components of the preserved Neolithic wells manifest the skills and the sophistication of their carpentry joints (i.e. tapping and lapping), which were undoubtedly also applied during the construction of the houses. The roof could be covered either with the split parts of logs or with bark or reeds or thatched straw [19].

Since then, the discovery of different elements such as bronze and steel has changed and improved the way wood is applied to building construction. Wattle and daub, a combination of woven wooden stripes and other adhesive materials, have been used to build walls for at least 6000 years.

A technique known as *timber framing*, combining the ductility and lightness of wood members and the compressive strength of stone blocks and bricks, has been found from the Bronze Age. Ulrich cites the archaeological discoveries in an early Italic building from c. 1800 BC and from the Iron Age (c. 1000 BC). A grid of timber beams was filled with thin panels made with woven twigs and mud, pieces of bark, and animal hides [20, 21].

The origin of half-timbered structures probably goes back to the Roman Empire, as in archaeological sites half-timbered houses were found and were referred to as Opus Craticium by Vitruvius [22]. The structure of the wall consists of a timber skeleton on a stone or brick plinth and of vertical uprights (*arectaria*) and horizontal members (*transversaria*). For the execution of the framework, Vitruvius recommends the *Abies alba* (silver fir) due to its strength, the high reach of the tree, its lightness, and its workability.

The Romans (50 AD) spread this mixed construction system throughout Gaul to the provinces of their jurisdiction, implementing the nonindigenous construction experience with their distinctively pragmatic standardized work in order to reduce the time of the construction. The traditional timber frame system made use of available and humble materials and was relatively simple to execute, without requiring specific preparation of the materials or skilled workmanship. Technical knowledge was handed down for a long time by master builders and craftsmen. See Fig. 8.7a.

Half-timbered constructions later spread not only throughout Europe, such as Portugal (*edificios pombalinos*), Italy (casa *baraccata*), Germany (*fachwerk*), Greece, France (*colombages* or *pan de bois*), Scandinavia, the United Kingdom (*half-timber*), and Spain (*entramados*) but also in India (*dhaji-dewari*) and Turkey (*himis*). By the Middle Ages (476–1500 AD), *timber framing* was reaching its heights with impressive structures such as the hammer-beam roof or the Westminster Hall, which now adjoins the British Houses of Parliament and was erected c. 1395 by Hugh Herland. See Fig. 8.7b.

In North America, a half-timbered construction tradition was brought from French and German immigrants. The most famous example is the Chicago Balloon frame, introduced by the carpenter George W. Snow in 1832 [24]. This method uses lighter timber elements than those used in the previous traditional buildings and

Fig. 8.7 From left to right. Timber framing technique (**a**) Opus craticum from Herculaneum, Italy (35–25 BC) [21]. (**b**) Westminster Hall, England (1395 AD) [23]

eliminates the diagonal elements as well as the traditional connections, which are substituted with simpler nailed connections.

Siegfried Giedion [25] explains that the invention of balloon framing in the second quarter of the nineteenth century was largely responsible for the phenomenal growth of American cities in the treeless regions west of Chicago. By using lumber sawed into standard-sized boards at mills in Chicago, Wisconsin, and Michigan from whence it was shipped throughout the West, relatively unskilled workmen were able to erect balloon-framed buildings both quickly and soundly. Balloon frames were given their stability by the machine-cut nails used to connect their sills, studs, joists, and plates. Their name, balloon frame, derives from the lightness of the frame construction as compared to the much heavier braced frames that preceded it. So successful was this new technology that nearly all wooden buildings erected today in the United States continue to employ a method of construction derived from nineteenth-century balloon framing (Fig. 8.8a).

Half-timbered construction has also been used in South America. The *quincha* (from the Quechua *qincha*, which means wall, fence, or enclosure), is a traditional construction system that consists mainly of a wooden structure with a cane or bamboo framework, covered with mud mixed with straw (Fig. 8.8b).

Quincha has very good thermal insulation due to the high thermal inertia provided by the layer of mud with which the rod is covered. It's the typical value of thermal conductivity is $\lambda = 0.56$ W/m K. And if the thickness of the panel is incremented with lightened soil mixtures or with the incorporation of additional layers of materials, both natural (rice husk or wood chips) and artificial (expanded polystyrene), it significantly improves its thermal conductivity performance and the value diminishes to 0.12 W/m K [27].

Another advantage is that it is a system composed of materials such as wood, earth, cane, and straw; it is totally breathable, generating a pleasant microclimate in the interiors it generates.

Fig. 8.8 From left to right. (**a**) Balloon framing in North America [24]. (**b**) Quincha in South America [26]

In Orient, there are also some outstanding examples of buildings built with timber. In China, temples are usually built with a timber frame on top of a stone base. The oldest wooden building in China is the Main Hall of the Nanchan Temple (Wutai), which dates back to 782 AD. The structure is symmetrically laid out, which is a typical structural feature of the Tang Dynasty (Fig. 8.9a). And Horyuji Temple (Nara) is the oldest wooden building in Japan. The five-story pagoda and the main hall were both originally built around the year 607 AD but after a fire were rebuilt in the year 711 AD. Fig. 8.9b shows the main hall (Kondo) and the five-story pagoda, whose central pillar is made from a tree that was cut in 594.

Large-scale fires in cities with many wooden buildings such as London (1666) and Chicago (1871) coupled with the intense dissemination of materials novelties such as steel and reinforced concrete and the development of engineering construction resulted in the progressive disuse of wood as the main constituent of the building structure, going to the background as part as a lot of roofs, mezzanines in some cases, and mainly used for finishes and equipment.

Nevertheless, technological advances, both in materials and constructive systems, with wood allow facing the design and construction of buildings of several modern floors adapted to contemporary spatial requirements, functionality, economy, security, and speed of execution. In the last decade of the twentieth century and this first decade of the twenty-first century, they have begun to design and build multistory multifamily housing buildings with structure and body of wood in some

Fig. 8.9 The oldest wooden buildings in Orient (**a**) Nanchan Temple in Wutai, China (782 AD) [28]. (**b**) Horyuji Temple in Nara, Japan. (607 AD rebuilt in 711 AD) [29]

cities of northern and central Europe, North America, and Oceania and in recent years in Japan.

8.4 Relevant Case Studies of Recently Raised Buildings

In recent years, a small renaissance has taken place regarding timber buildings, and the use of wood is again being perceived as a structural and cladding material. New materials have been produced derived from wood such as the well-known chipboards and elements of laminated wood, more recently oriented strand board (OSB), laminated veneer lumber (LVL), cross-laminated wood, beams, joists, and composite and precast modules of mezzanines and roofs.

Together, with the new technical presentations of wood and the materials derived from it, prefabricated construction systems have been created, and industrialized products that respond to the advanced benefits of these innovations have been developed as the novel prefabricated systems made to measure with cross-laminated wood and the latest post-tensioned wood. Also, the increased popularity of mass timber is allowing wood buildings to stretch new heights beyond traditional wood-frame construction, ushering in an era of high-rise wood buildings.

Pritzker prize laureate in 2014, Shigeru Ban, is unique in this regard of timber as a construction material, inspired by the architectural tradition of his native Japan, Ban one of the most famous architects working in wood today.

Ban pays attention to environmental factors in his designs from the smallest to the largest scales, including infrastructure, natural sources, local materials, and climate. The principal objective of Ban's design approach is satisfying the needs of current and future generations. Instead of the high-tech designs that are the usual response to today's architectural sensibility, he adopted an easily accessible, economical, and environmentally sensitive design concept.

For the Japanese Pavilion in the environment-oriented expo in Hannover, Germany (2000) Shigeru Ban's environmentally sensitive architectural design approach included these major items:

Concept: The pavilion is made up of floor texture and top covering. At the base of the construction, reusable steel boxes filled with sand were used instead of concrete in an attempt to reduce construction waste.

Structure: Ban collaborated with Frei Otto and Buro Happold to develop the roof's grid shell system that reduces costs by minimizing the use of wooden joints. Paper tubes, 40 m in length with a radius of 12 m, were used in the tunnel arch and connected by laminated wood and polyester fabric tape. Ban disassembled the structure of his system without using vertical conveyor elements. The outer shell and top covering are easily removable and reusable. See Fig. 8.10.

Roof: It was made of a fire- and water-resistant, translucent membrane. Along with ecological materials, Ban used the traditional Japanese architectural understanding of *shoji*, to benefit from natural light and save energy. See Fig. 8.11.

Shoji is a kind of door typical of Japanese homes, also known as screens, which consist of a sheet of translucent *washi* paper framed in wood, which is generally sliding or folded into different panels to occupy the minimum space. Its use dates

Fig. 8.10 Shigeru Ban's Japanese Pavilion in the Expo in Hannover, Germany. From left to right. (**a**) Floor and top covering architectural concept. (**b**) Detail of the arched wooden structure [30]

Fig. 8.11 Adaptation of traditional Japanese architecture to benefit from natural light and save energy. From left to right. (**a**) Pavilion's roof system [30]. (**b**) Shoji system [31]

Fig. 8.12 Aspen Art Museum by Shigeru Ban. (**a**) Gridwork of wood plies interlaced to create an outer screen. (**b**) The glass enclosure above the truss structure produces a luminous and dynamic patterning [32]

back to the Han dynasty in China, although they began to form part of the Japanese tradition from the fifteenth century.

Ban's sustainable vision is present in most of his projects. Another example is the Aspen Art Museum (2014) that has a ceiling entirely composed of a wooden truss system. For the ceiling, the wood pieces merely needed to be cut into the appropriate shape, laminated, and fastened together. To attempt the same structure with any other material would demand more labor and more room for error. See Fig. 8.12.

And his Tamedia office building built in 2013 in Zurich, Switzerland, is even a more extreme example. A seven-story pin joint structure made almost entirely of prefabricated wood pieces. The design requires a high degree of precision in both material production and construction. These unfastened wooden pieces can be replaced independently and with relative ease at least compared to a concrete frame of similar design. There is the matter of aesthetics to consider as well. Ban's favoritism for wood also springs from an appreciation for its color, texture, and visual malleability [33] (See Fig. 8.13).

Fig. 8.13 Tamedia Office Building by Shigeru Ban. From left to right. (a) Joint Detail and (b) Interior view of the stairs [33]

8.5 Future Approach to Sustainable Buildings

Timber is among the few materials on the market today that is easily sustainable. When taken from a forest that is sustainably managed, the supply can be considered indefinite. The process by which timber is harvested and manufactured is also relatively ecological. It is recyclable, reusable, and naturally renewable.

Moreover, its excellent thermal, insulating, and resistant properties make it useful for different kinds of applications in buildings, ranging from structural beams and frames, insulating envelopes, windows, and door frames to wall and flooring materials and furniture.

The importance of wood not only as a building material but also in the elaboration of numerous industrial products has required a great variety of transformations that have opened an important field of applications and uses for this material obtained from trees.

How should one use timber? While there are limitless possible designs and construction is based on both engineering and cultural practice, timber has a high strength to weight ratio and is used most efficiently in structures where it is carrying a lot of its own self-weight.

Sustainably grown and harvested wood has a smaller carbon footprint than concrete and steel, making it a good choice for even large buildings. A mass timber building's carbon footprint is estimated to be almost 75% less than a concrete and steel building of similar size [34].

Also, the Consortium for Research on Renewable Industrial Materials (CORRIM) has found that in life cycle analyses—the measurement of how much energy a material requires, from harvesting it to disposing of it after it is no longer needed—wood outperforms steel by 17% and concrete by 16% [33].

Timber has been present in our cultural heritage for more than 12,000 years. And it is still a valid sustainable resource. Outstanding examples in contemporary architecture demonstrate its unique characteristics. At the smallest scale, the elements that make up the structure of wood contribute to its various properties at a macro scale.

At the largest scale, the knowledge of material properties increases the already significant potential for using timber in the built environment. Bringing these two scales together provides directions for future research that will shape environmental outcomes highlighting the importance of wood and timber in sustainable building.

References

1. Thomas, P. A. (2014). *Trees: Their natural history*. London: Cambridge University Press.
2. Zhang, J., Nieminen, K., Alonso Serra, J. A., & Helariutta, Y. (2014). The formation of wood and its control. *Current Opinion in Plant Biology, 17*(2014), 56–63.
3. Rowell, R. M. (1984). *The chemistry of solid wood. Advances in chemistry*. Washington, DC: American Chemical Society.
4. Ramage, M. H., Burridge, H., Busse-Wicher, M., Fereday, G., Reynolds, T., Shah, D. U., Wud, G., Yuc, L., Fleming, P., Densley-Tingleye, D., Allwoode, J., Dupreec, P., Lindenb, P. F., & Schermane, O. (2017). The wood from the trees: The use of timber in construction. *Renewable and Sustainable Energy Reviews, 68*(2017), 333–359.
5. Asdrubali, F., Ferracuti, B., Lombardi, L., Guattari, C., Evangelisti, L., & Grazieschi, G. (2017). A review of structural, thermo-physical, acoustical, and environmental properties of wooden materials for building applications. *Building and Environment, 114*(2017), 307–332.
6. Pingoud, K., Skog, K. E., Martino, D., Tonosaki, M., & Xiaoquan, Z. (2006). Chapter 12: Harvested wood products. In *IPCC guidelines for national greenhouse gas inventories*. Hayama: Institute for Global Environmental Strategies.
7. Bamaca Figueroa, E., Kanninen, M., Louman, B., Pedroni, L., et al. (2004). *Contenido del carbono en los productos y residuos forestales generados por el aprovechamiento y el aserrio en la reserva de Biosfera Maya*. Center for International Forestry Research.
8. Wiedenhoeft, A. (2010). Structure and function of wood. In R. J. Ross (Ed.), *Wood handbook: Wood as an engineering material*. Centennial ed. (General technical report FPL GTR-190). Madison: U.S. Department of Agriculture, Forest Service, Forest Products Laboratory.
9. Batchelor, W. J., Conn, A. B., & Barker, I. H. (1997). Measuring the fibril angle of fibres using confocal microscopy. *Appita Journal, 50*(1997), 377–380.
10. Huang, C.-L., Lindstrom, H., Nakada, R., & Ralston, J. (2003). Cell wall structure and wood properties determined by acoustics—A selective review. *Holz als Roh- und Werkst, 61*, 321–335.
11. Cave, I. D., & Walker, J. F. C. (1994). Stiffness of wood in far own plantation softwood: The influence of microfibril angle. *Forest Products Journal, 44*(1994), 43–48.
12. Dinwoodie, J. M. (2000). *Timber: Its nature and behaviour*. Abingdon: Taylor and Francis.
13. BSI. (2014). BS EN 1992-1-1:2004. Eurocode 2. Design of concrete structures.
14. BSI. (2009). BS EN 1993-1-1:2005. Eurocode 3. Design of steel structures.
15. BSI. (2009). BS EN 338:2009. Structural timber. Strength classes.
16. Archeological Research Services. (2019). *Stunning mesolithic discovery*. Retrieved from https://www.archaeologicalresearchservices.com/2019/10/25/stunning-mesolithic-discovery/.
17. Archeology News Network. (2013). *10,000 year old mesolithic hut recreated*. Retrieved from https://archaeologynewsnetwork.blogspot.com/2013/07/10000-year-old-mesolithic-hut-recreated.html.
18. Brunskill, A. (2019). Mesolithic structure with surviving timbers found at Killerby Quarry. *Current Archeology 4*, 2019.

19. Archeologické 3D virtuální muzeum. (2020). *The people from Longhouses*. Retrieved from http://www.archaeo3d.com/en/lide-z-dlouhych-domu/lide-z-dlouhych-domu/lide-z-dlouhych-domu/.
20. Ulrich, R. B. (2007). *Roman woodworking*. New Haven: Yale University Press.
21. Stellacci, S.. & Rato, V. (2019). Timber-framing construction in herculaneum archaeological site: Characterisation and main reasons for its diffusion, *International Journal of Architectural Heritage*, https://doi.org/10.1080/15583058.2019.1672827
22. Langenbach, R. (2009). *Don't tear it down: Preserving the earthquake resistant vernacular architecture of Kashmir*. New Dehli: UNESCO.
23. Waddell, G. (1999). The design of the Westminster Hall Roof. *Architectural History, 42*, 47–67.
24. Sprague, P. E. (1981). The origin of balloon framing. *Journal of the Society of Architectural Historians, 40*(4), 311–319. University of California Press.
25. Giedion, S. (1941). *space, time and architecture*. Cambridge: MIT Press.
26. Galvez, H. R. (2020). *Quincha una tradición de futuro*. Retrieved from http://www.mimbrea.com/quincha-una-tradicion-de-futuro/.
27. Wieser, M., Onnis, S., & Meli, G. (2018). Conductividad térmica de la tierra alivianada con fibras naturales en paneles de quincha. In *18 Seminario Iberoamericano de Arquitectura y Construcción en Tierra*, La Antigua, Guatemala.
28. China Cultural Heritage. (2020). *Main hall of Nanchan Temple*. Retrieved from http://www.china.org.cn/.
29. Woodard, A. C., & Milner, H. R. (2016). Chapter 7—Sustainability of timber and wood in construction. In *Sustainability of construction materials (second edition)* (Woodhead publishing series in civil and structural engineering) (pp. 129–157). Sawston: Woodhead Publishing.
30. Bulut, D. M., & Gürani, F. Y. (2018). A study of Shigeru Ban's environmentally sensitive architectural design approach. *Gazi University Journal of Science, 6–3*(2018), 147–157.
31. La Cuisine International. (2017). *Shoji: una milenaria herramienta para la nueva arquitectura*. Retrieved from https://www.lacuisineinternational.com/es/noticias/diseno-y-tendencias/shoji-una-milenaria-herramienta-para-la-nueva-arquitectura/.
32. David. (2014). *First look at the completed Aspen Art Museum by Shigeru Ban*. Retrieved from https://www.designboom.com/architecture/aspen-art-museum-shigeru-ban-08-06-2014/.
33. Walker, C. (2014). *Material masters: Shigeru Ban's work with wood*. Retrieved from https://www.archdaily.com/573818/material-masters-shigeru-ban-s-work-with-wood.
34. Killough, D. (2015). *Building material of the future—wood?* Green Building Elements. Retrieved from https://greenbuildingelements.com/2015/03/05/building-material-future-wood/.

Chapter 9
Laminated Timber Buildings: An Overview of Environmental Impacts

Rahman Azari and Maryam Singery

9.1 Introduction

Buildings contribute significantly to the adverse impacts on the environment as well as energy, material, and water consumption. In the United States, buildings' operations (i.e., heating, cooling, lighting, and operating appliances and equipment) make up 39% of annual energy use, 74% of electricity consumption, and 40% of CO_2 emissions [1]. In 2018, 600 million tons of construction and demolition waste were produced, 90% of which was due to demolition and 10% due to construction [2]. Also, 6% of manufacturing energy is used to produce construction materials [3]. To achieve the American Institute of Architects' goal of "net-zero emissions in the building sector by 2050," significant changes must happen in the way buildings are designed and constructed. One major development in response to this call is the timber revival in the tall construction industry. In the past few years, a new generation of super-high-performance tall buildings has been developing that attempts not only reduce not only the building operational energy use but also its embodied energy content and carbon emissions by using timber-framed structures. An example is the mixed-use all-timber "Mjøstårnet" building (18 floors, 11,300 m^2, 85.4-m tall) constructed in 2019 in Brumunddal, Norway, which uses glulam and

R. Azari
Department of Architecture, Pennsylvania State University, University Park, Pennsylvania, USA
e-mail: razari@psu.edu

M. Singery (✉)
Architectural & Engineering Design Technology Program, South Texas College, McAllen, Texas, USA

A. Sayigh (ed.), *The Importance of Wood and Timber in Sustainable Buildings*, Innovative Renewable Energy, https://doi.org/10.1007/978-3-030-71700-1_9

cross-laminated timber to support horizontal and vertical structural loads.[1] As of December 2020, the building bears the title of the world's tallest timber building [4].

The history of wood and timber buildings goes back to the Neolithic times in thousands of years ago in areas such as Europe and ancient Japan [5], continues to the ancient Mediterranean and Roman timber-framed *opus craticum* housing typology mentioned by Vitruvius in his "Ten Books on Architecture" [6], and extends to the medieval and modern times in different countries. In Asia, the fully wooden 67-m-tall Sakyamuni Pagoda of Fogong Temple built in China in 1056 is one of the oldest examples of timber skyscrapers. In the United States, the European settlers used wood as a faster method of construction. In the 1830s, the Americans developed balloon framing, as a faster and less expensive technique for house construction, compared with timber construction. Similar to how the Great Fire of London in 1666 facilitated a change from wood to masonry construction, the urban fires of the late nineteenth century in the United States, along with timber's limitation with regard to fire, safety, and tall construction, and more importantly the technological opportunities presented by new materials, all led to tall buildings that were now built with steel and later concrete. Despite the developments in tall building construction in the late nineteenth and twentieth century, wood continued as a major construction material in low-rise buildings in the United States.

In recent years, timber has received renewed interest worldwide as a construction material for high-rise building applications, because of four critical factors including (a) concerns about the environmental impacts of steel and concrete buildings; (b) timber's potentials to lower the carbon footprint and embodied environmental impacts of tall buildings; (c) the enhancement of timber's mechanical properties, structural performance, and safety factors through wood products such as Cross-Laminated Timber (CLT) and Glue-Laminated Timber (GLT); and (d) timber's aesthetic qualities. Architects and engineers have thus begun to understand timber as an efficient material, from both structural and environmental perspectives, that could replace steel and concrete in tall buildings. However, the environmental consequences of the massive application of timber in building structures beyond current levels need to be further examined. Cornwall [7] suggests that mass replacement of structural steel with wood would require 40% of forest growth per year globally which would lead to forest degradation.

9.2 Forests and Deforestation

Considering that about one-third of softwood lumber and structural panel consumption in the United States is due to new housing construction which is expected to grow [8], wider application of timber in buildings would translate to effects on forests, greater rates of deforestation, and subsequent effects on carbon cycles and embodied environmental impacts.

[1] Architectural Digest. March 22, 2019. https://www.architecturaldigest.com/story/worlds-tallest-timber-framed-building-finally-opens-doors

Forests comprise 31% of the global land area, and more than 53% of the global forests are located in five countries including Russia (20.1%), Brazil (12.2%), Canada (8.5%), the US (7.6%), and China (5.4%) [9]. Forests are carbon pools that continuously exchange carbon with the atmosphere, driven by two forces including (a) biogeochemical natural processes such as photosynthesis, respiration, decomposition, and disturbances (fires, pest infestations, etc.) and (b) anthropogenic activities such as tree harvesting, thinning, replanting [10]. The complex contribution of forests to carbon cycles and climate change can be examined at four levels:

- *Forests are carbon pools*. In other words, they act as massive carbon reservoirs with the capacity to store or emit carbon.
- *Forests are CO_2 sinks*. That is, they pull CO_2 out of the atmosphere and store carbon in plants and soil. More specifically, trees and vegetations in forests use the photosynthesis process to absorb CO_2 from the air and combine it with sunlight and water to make glucose and oxygen. The carbon from CO_2 is then stored (i.e., sequestered) in the plant biomass (aboveground, belowground, deadwood, litter) and soil organic matter. About 50% of the tree biomass (in its dry form) is carbon [11]. The harvested wood continues to store carbon. Despite the slight decrease in the total area of the U.S. forests, the estimates show that the total carbon stock of the U.S. forest areas, including forest ecosystem and harvested wood, has increased in 2019 by 9.9% to 58,720 million metric tons (MMT) of carbon,[2] as compared with 1990 figures [12]. This increase in carbon stocks is believed to be primarily due to forest management practices leading to the increased growth rate of the aboveground biomass [12].
- *Forests are CO_2 sources*. In other words, forests lose part of their carbon stock by emitting CO_2. This is done through both human actions (e.g., harvesting, thinning) and natural activities such as plant respiration, decomposition, and disturbances (e.g., fire, drought).
- *Forests are NET carbon sinks*. Accounting for both CO_2 emissions and removals in a year, the forests in the U.S. and North America are net carbon sinks as they remove more CO_2 than they emit [10, 12]. In other words, the uptake of carbon by forests in the photosynthesis process exceeds the release of carbon by harvesting, respiration, decomposition, or fires; therefore, these forests act as a net carbon sink [10]. The net CO_2 removal by the U.S. forest lands is estimated to be 663.2 (MMT CO_2-eq), equivalent to 9.95% of the annual CO_2 emissions in the country. The carbon sink of the North American forests is estimated to have offset the anthropogenic CO_2 emissions by 16–52% in the past two decades [10]. However, there are concerns that anthropogenic global warming and natural disturbances may lead to forests becoming a weaker carbon sink or even a net carbon source by the end of the century [13].

Considering the benefits of forests, deforestation leads to significant and complex environmental consequences. Deforestation and clearing of forests are largely done by human agents to meet the land, energy, food, and wood demand of the growing global population. The major direct causes include the expansion of land

[2] Carbon constitutes 27.24% of the CO_2 mass. One ton of carbon translates to 3.67 ton of CO_2.

for agriculture, food production, or soya bean and palm oil plantation, logging to address demand and make financial profits, oil extraction and mining, construction projects, and natural causes such as fire catastrophes, droughts, hurricanes [14]. Deforestation, however, leads to changes in the exchanges of energy and water between land and the atmosphere and is considered a major source of the CO_2 emissions caused by land use and associated changes and a contributor to global warming due to its CO_2 release [15]. Deforestation also affects climatic and precipitation patterns, the hydrological cycle, surface albedo, and surface temperature, increases vulnerability to fire catastrophes, and decreases atmospheric water vapor content due to lesser evapotranspiration [15]. The changes in regional surface temperature in turn directly affect the energy needed for the heating and cooling of buildings. In addition, deforestation leads to biodiversity loss by affecting the population and species of trees, plants, animals, and microorganisms living in forests. The extent to which deforestation affects biodiversity, however, is varied depending on factors such as geographical location, climate, soil type, and forest type [9]. Maintenance of biodiversity, land clearing for agriculture and food production, and illegal harvesting for profit are some of the main considerations in harvesting [16].

While the global rate of deforestation is alarming, its effects are compensated by afforestation, reforestation, and sustainable forestry and land management, all driven by factors such as wood demand, maintenance of biodiversity, and environmental protection. In addition, sustainable harvesting facilitates the optimum and quicker growth of trees by providing them with physical space and lesser competition for sunlight and nutrients. Sustainable harvesting and using timber buildings with long life spans allow for more carbon sequestration by the timber than what can be achieved by an unmanaged forest [17]. Because of afforestation and sustainable harvesting, the net loss of forest areas has been determined to be lesser than the rate of deforestation, and this area loss is also decreasing in absolute value [9].

9.3 Timber in Buildings, Benefits, and Limitations

Wood is the fibrous, porous, and organic tissue of trees and vegetation. Once the tree is harvested, the wood present in its trunk is shaped and processed into the timber. Different types of processing that are applied on harvested wood include (a) drying to improve mechanical properties [17], (b) dimensional processing to improve dimensional stability and structural performance [17], and (c) wood treatment to improve resistance to biological degradation, thermal stability and fire resistance, UV resistance, and mechanical properties [17, 18].

Wood is a material with numerous advantages. It is a natural material with high potentials to be reused and recycled. It is also biodegradable and renewable with short forestry rotation periods that are close to buildings' life span [17]. In addition, significant forest resources globally, combined with sustainable forestry and land management, ensure the sufficient long-term supply of timber for various applications including in buildings.

On the other hand, wood is flammable, and fire safety is an important consideration in the design of timber buildings. Additionally, wood is a material with unpredictable behavior due to the presence of twists and knots [7]. Wood is also susceptible to the temperature and moisture of its surrounding environment.

Carbon

The timber harvested from the forests that are sustainably managed is arguably one of the most sustainable materials and is considered carbon-positive [16]. Wood and timber contribute to the carbon sinks through three key pathways, as shown in Fig. 9.1:

- ***Carbon removal***: Wood in the standing forest removes CO_2 from the atmosphere through the photosynthesis process. At the age of 10, which is the most productive stage for a young tree from the carbon storage perspective, the tree removes 48 pounds of CO_2 per year [19].
- ***Carbon storage***: About 50% of the dry wood content is carbon [11], and the carbon stored in wood is released into the atmosphere when the wood decomposes or burns. So, by converting wood into timber and other wood-based product, the carbon is stored for the life span of timber. The quantity of carbon storage depends on the tree species as well as the geographical latitude and location of the forest from which the tree is harvested [17]. As a rule of thumb, there is 1.84 kg of equivalent carbon stored per kg of harvested over-dried wood [20]. The carbon stored in the cross-laminated timber (CLT) is considered to be 585 kg of CO_2-eq (i.e., 159 kg of carbon) per cubic meter by Nakano et al. [21] and 700 kg of CO_2-eq (i.e., 190 kg of carbon) per cubic meter by Chen et al. [22]. Including carbon storage in the estimation of cradle-to-site CO_2 emissions of CLT timber buildings is reported to reduce the emissions from 32% [23] to 48% [22]. It should be noted that the carbon remains stored in the wood products until they are burned or decomposed.

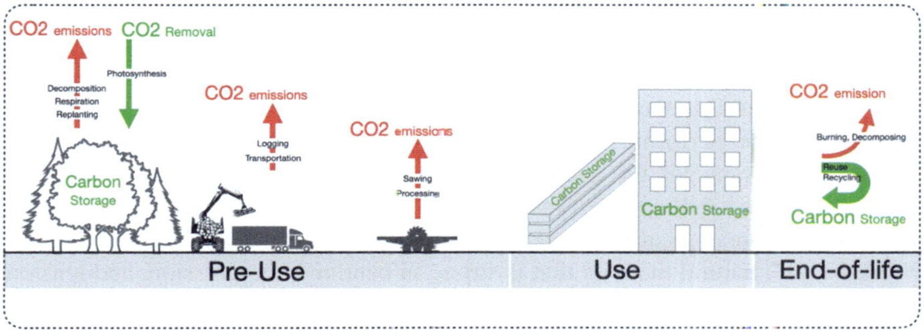

Fig. 9.1 Carbon cycles of timber products in buildings

- ***Carbon avoidance***: CO_2 emissions, as well as other embodied emissions, are avoided primarily at the production stage and end-of-life state by replacing the energy- and carbon-intensive steel and concrete structure in low-, mid-, and high-rise buildings with timber. A building made from wood would is estimated to a carbon footprint equivalent to one-third of steel or concrete buildings [7].

On the other hand, wood and timber also contribute to the CO_2 sources through the following two pathways:

(a) *Carbon emission in the forest*: As mentioned before, trees emit CO_2 emissions through respiration, decomposition, and replanting.
(b) *Carbon emission during production*: Carbon is emitted also during logging, sawing, processing, and treatment of timber and other wood-based products. Compared with other materials such as concrete or steel, timber's production needs lower levels of technology and is a process that consumes lesser energy and emits lesser carbon [16], which translates into timber's carbon avoidance potentials. Carbon emitted by the production of framing lumber, for instance, equals 11.3% of that emitted by the production of concrete, 15% of that emitted by the production of recycled (100%) steel, and 4.7% of virgin steel of equal weight [24]. Comparing the carbon footprint of materials based on equal weight is of course not a fair comparison as different quantities of materials would be needed to serve the same function. Most of the energy used for timber production is used for drying. Robertson et al. [25] compared a mid-rise office building structure and enclosure built with reinforced concrete with an alternative laminated timber design and showed that the global warming potential of timber building including its carbon storage effect was 30% that of reinforced concrete.

Mechanical Properties and Strength

The mechanical and strength properties of wood (such as strength, stiffness) depend on factors such as the moisture content, density, type, and species of wood, specific gravity, temperature, the slope of grain, and the presence and frequency of knots [24]. Moisture content is a critical factor. There is an inverse relationship between the wood's strength and stiffness and its moisture content up to a certain point called fiber saturation point (FSP), above which the wood's mechanical properties do not vary with moisture content [17]. Therefore, drying is a necessary step to achieve the desired structural performance for timbers used in a structural application. Drying offers other advantages too such as increasing thermal resistance of timber and lesser weight for handling and transportation. Structural timber is dried to achieve a 12–20% moisture content [17].

Timber is a natural material that is strong in bending, compression, and tension [16], with high strength-to-weight and modulus-to-weight ratios similar to steel and higher than concrete, which makes timber efficient for tall building structural applications where the self-weight of the structure constitutes a great portion of the loads to resist [17]. One kilogram of the Douglas Fir tree, for instance, is 3.5 times stronger than steel of equal weight [7].

Thermal Resistance and Heat Capacity

Thermal resistance (*R*-value) of wood is considerably higher than that of metals but 25% to 50% lesser than common insulations [24]. Moisture content is one of the factors affecting thermal resistance where the resistance increases as the moisture content decreases. Along with temperature, the moisture content also affects the wood's heat capacity, which is practically independent of species or density of wood [24]. Heat capacity increases with moisture content. According to the Canadian Wood Council, wood is 10 times better than concrete and 400 times better than steel in thermal resistance, which translates into lesser insulation needed for wood framing. The heat capacity and volumetric heat capacity of wood and concrete are about 47% and 11%, respectively. The thermal performance of timber is limited mainly because it cannot absorb and release heat fast enough to address the day-night thermal cycles [17].

9.4 Environmental Life Cycle Impacts

The environmental life cycle impacts of buildings are often assessed with the environmental life cycle assessment (LCA) methodology. The methodology consists of four steps including goal and scope definition, inventory modeling, impact assessment, and interpretation of results. Through these four steps, the contribution of buildings, products, or materials of interest to the ecological health and consumption of resources is assessed for the entire or part of the life cycle. More specifically, different types of energy, materials and other resources, emissions, and waste that are used/produced/emitted during various stages of the life cycle from raw material extraction, manufacturing, and construction to maintenance, disassembly, and demolition, as well associated transportation, are tracked, quantified, classified, and aggregated into various categories of environmental impacts such as global warming, acidification, eutrophication, smog formation, cumulative energy use, and ecotoxicity. Figure 9.2 illustrates a schematic of the LCA process. Because a key purpose of using the LCA methodology is to compare different alternatives of materials, processes, products, systems, or buildings, it is essential for a fair comparison to define a functional unit based on which these alternatives are compared based on doing the same function, rather than based on equal mass, weight, quantities, and similar. For example, when the LCA's scope is comparing material alternatives for the structure of a building, the functional unit could be for the materials to resist the same predefined live or dead loads in a certain structural system.

The LCA studies on timber buildings report diverse results. The reported results are usually based on the belowground reinforced concrete foundation and aboveground laminated timber structure. The studies that report the energy-related effects of the operation stage of building life span, such as Guo et al. [26], are rare. Most studies instead report the cradle-to-site results. The results, however, are difficult to compare due to insufficiently or inconsistently reported assumptions and specifics. Comparing timber and reinforced concrete buildings using LCA, Skullestada et al.

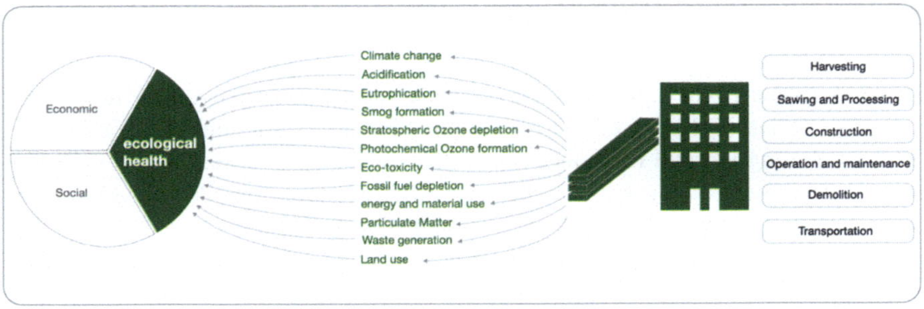

Fig. 9.2 LCA methodology can assess the environmental impacts of a laminated timber building over its complete or partial life cycle

[27] suggest that higher savings in climate change impacts are possible in timber tall buildings versus concrete tall buildings, as compared with the savings that are achieved in low- to mid-rise buildings by replacing reinforced concrete with timber structure. More specifically, Skullestada et al. [27] reported climate change impacts of timber buildings to range from 140–235 kg CO_2-eq per square meter, depending on building height, which could lead to 34% to 84% less climate change impacts (i.e., 246–634 kg CO_2-eq of GHG emission savings per square meter), depending on production technology and building height [27]. Robertson et al. [25] used LCA methodology to compare a laminated timber structure plus enclosure alternative to an existing reinforced concrete five-story office building. This study that included cradle-to-site as well as building maintenance stages reports an almost identical embodied energy content for two building types due to the manufacturing stage but estimates that the laminated timber building has a global warming potential (GWP) equal to 30% of the concrete building's [25].

Katsuyuki et al. [23] conducted a cradle-to-site LCA study on a two-story research and development CLT building in Japan and reported 711 kg of CO_2-eq GWP per square meter. Including the carbon storage of CLT components reduces the emissions to 485 kg of CO_2-eq per square meter. In this study, 53% of the building's entire GHG emissions were found to be associated with the construction of reinforced concrete foundation while CLT manufacturing was estimated to account for 17% of the emissions. Chen et al. [22] conducted a whole-building LCA by including all stages of the building life cycle including beyond end-of-life to compare an American 12-story laminated timber mixed-use building with its reinforced concrete equivalent. This study reported lesser environmental impacts of laminated timber buildings in several categories including global warming potential (by 21%), eutrophication potential (by 29%), and Ozone depletion potential (by 8%). On the other hand, timber building was estimated to have higher impacts on smog formation potential (by 19%), acidification potential (by 12%), and human health (by 6%). Laminated timber building's cradle-to-gate CO_2 emissions were estimated to be 21% lesser than a concrete building, which could increase to 69.5% when carbon storage is accounted for.

Table 9.1 lists some examples of LCA research on timber buildings along with their scope and system boundaries

Table 9.1 Energy use and carbon emissions of laminated timber buildings compared with other buildings, as reported by selected studies

	Context	Building type	Floors	Energy consumed (MJ/m^2)				CO_2-eq emissions (kg/m^2)			
				Pre-use	Use	Post-use	Total	Pre-use	Use	Post-use	Total
Robertson et al. [25]	Mid-rise office (5-storey, 50-years), West Canadian, cradle-to-site plus maintenance[a]	Reinforced concrete structure and steel stud	5	3510 4600		NR	3510 4600	420		NR	420
		Above ground laminated timber structure and enclosure	5	3490 8170		NR	3490 8170	126		NR	126
Guo et al. [26]	Residential buildings, Harbin China, 50-year, severe cold climate[b]	Reinforced concrete building	4	1541.1	34,314.1	90.0	35,945.2	308.2	6916.7	18.0	7242.9
			7	1463.2	33,320.5	90.0	34,873.7	292.6	6698.7	18.0	7009.3
			11	1350.2	32,839.7	90.0	34,279.9	270.0	6596.2	18.0	6884.2
			17	1326.6	32,538.51	90.0	33,955.11	265.3	6532.1	18.0	6815.4
		CLT timber building	4	847.5	31,455.1	18.0	32,320.6	−84.0	6314.1	45.6	6275.7
			7	790.8	30,621.8	18.0	31,430.6	−85.7	6121.8	46.5	6082.6
			11	711.3	30,160.2	18.0	30,889.5	−96.9	6025.6	52.0	5980.7
			17	694.1	29,884.6	18.0	30,596.7	−97.4	5967.9	52.3	5922.8
Chen et al. [22]	The mixed-use building, United States, 60-years	Reinforced concrete building	12					238	4.09 + 2799	14.71	257.95 + 2799 = 3056.95
		Laminated timber building	12	NR	NR	NR	NR	183	7.09 + 2799	14.47	204.54–125.59[c] + 2799 = 2877.95
Katsuyuki et al. [23]	Research building in Japan	Reinforced concrete foundation with timber structure	2	NR	NR	NR	NR	711	NR	NR	711–226[a] = 485

NR not reported
[a]Feedstock plus product energy
[b]Maintenance not included
[c]Carbon storage

9.5 Conclusion

While the application of timber in buildings has a long history, its application in the multistory building was affected by developments of steel and reinforced concrete in the nineteenth century, along with concerns regarding the safety, fire, and structural performance of timber. However, the advent of engineered timber products in the past few decades has opened new possibilities to bring back timber as a structural material in tall buildings. Examples of such products include cross-laminated timber (CLT) and glue-laminated timber (GLT) that offer enhanced mechanical properties, structural capacity, fire resistance, and durability. These developments, along with the perception of timber having a lesser carbon footprint, have contributed to renewed interest in multi-story buildings built with laminated timber.

The environmental impacts of timber buildings, however, are not easy to assess or generalize due to the complex effects of harvesting on deforestation and biodiversity loss or significant effects of system boundaries issues on the assessment of environmental impacts. For example, Woodard and Milner [16] suggest that the LCA of wood-based products requires the inclusion of plantation, fertilization, thinning, and management practices in system boundaries. In addition, while laminated timber buildings have superior carbon storage performance and lesser CO_2 emissions, there is research [22] that shows they can lead to higher smog formation potential, acidification potential, or great effects on human health, as compared with reinforced concrete buildings.

Future research also needs to address the issues of uncertainty in quantifying the embodied carbon emissions and impacts of timber buildings.

References

1. EIA. (2019). *Energy consumption by sector*. Retrieved from US Energy Information Administration: https://www.eia.gov/totalenergy/data/monthly/pdf/sec2_3.pdf
2. EPA. (2020a). *Advancing sustainable materials management: 2018 fact sheet*. US Environmental Protection Agency. Retrieved from https://www.epa.gov/sites/production/files/2020-11/documents/2018_ff_fact_sheet.pdf
3. EIA. (2020). *Annual energy outlook*. Retrieved from https://www.eia.gov/energyexplained/use-of-energy/industry.php
4. Guinness. (2020, November). *Tallest wooden building*. Retrieved from Guinness World Records: https://guinnessworldrecords.com/world-records/79569-tallest-wooden-building/
5. Williams, J. H. (1971). Roman building-materials in south-East England. *Britannia, 2*, 166–195.
6. Langenbach, R. (2007). From "opus Craticium" to the "Chicago frame": Earthquake-resistant traditional construction. *International Journal of Architectural Heritage, 1*, 29–59.
7. Cornwall, W. (2016). Tall timber. *Science, 353*(6306), 1354–1356.
8. Howard, J. L., & Liang, S. (2019). *U.S. timber production, trade, consumption, and price statistics, 1965–2017*. Washington, DC: United States Department of Agriculture.
9. FAO. (2020). *The state of the World's forests 2020*. New York: FAO and UNEP.
10. USGCRP. (2018). *Second state of the carbon cycle report*. Washington, DC: US Global Change Research Program.

11. Battles, J. J., Bell, D. M., Kennedy, R. E., Saah, D. S., Collins, B. M., York, R. A., et al. (2018). Innovations in measuring and managing forest carbon stocks in California. In *California's Fourth Climate Change Assessment*.
12. EPA. (2020b). *U.S. Inventory of Greenhouse gas emissions and sinks*. Washington, DC: United States Environmental Protection Agency.
13. Sitch, S., Huntingford, C., Gedney, N., Levy, P., Lomas, M. R., Piao, S. L., et al. (2008). Evaluation of the terrestrial carbon cycle, future plant geography and climate-carbon cycle feedbacks using 5 dynamic global vegetation models (DGVMs). *Global Change Biology, 14*, 1–25.
14. Tejaswi, G. (2007). *Manual on deforestation, degradation, and fragmentation using remote sensing and GIS*. Rome: FAO.
15. IPCC. (2020). *Climate change and land*. Geneva: Intergovernmental Panel on Climate Change.
16. Woodard, A., & Milner, H. (2016). Sustainability of timber and wood 7 in construction. In J. Khatib (Ed.), *Sustainability of construction materials* (pp. 129–157). Cambridge: Woodhead.
17. Ramagea, M. H., Burridge, H., Busse-Wicher, M., Fereday, G., Reynolds, T., Shah, D. U., et al. (2017). The wood from the trees: The use of timber in construction. *Renewable and Sustainable Energy Reviews, 68*, 333–359.
18. Rowell, R. (2007). Chapter 22. Chemical modification of wood. In F. Stoyko & B. Debes (Eds.), *Handbook of engineering biopolymers—homopolymers, blends and composites*. Munich: Hanser.
19. UFN. (2019). *Trees improve our air quality*. Retrieved from Urban Forestry Network: http://urbanforestrynetwork.org/benefits/air%20quality.htm
20. Lippiatt, B. (2007). *BEES 4.0: Building for environmental and economic sustainability, technical manual and user guide*. Gaithersburg, MD: National Institute of Standards and Technology.
21. Nakano, K., Koike, W., Yamagishi, K., & Hattori, N. (2020). Environmental impacts of cross-laminated timber production in Japan. *Clean Technologies and Environmental Policy, 22*, 2193–2205.
22. Chen, Z., Gu, H., Bergman, R. D., & Liang, S. (2020). Comparative life-cycle assessment of a high-rise mass timber building with an equivalent reinforced concrete alternative using the Athena impact estimator for buildings. *Sustainability, 12*(4708), 1–15.
23. Katsuyuki, N., Karube, M., & Hattori, N. (2020). Environmental impacts of building construction using cross-laminated timber panel construction method: A case of the research building in Kyushu, Japan. *Sustainability, 12*(2220), 1–14.
24. FPL. (2010). *Wood handbook; wood as an engineering material*. Madison, WI: Forest Products Laboratory. United States Department of Agriculture Forest Service.
25. Robertson, A. B., Lam, F. C., & Cole, R. J. (2012). A comparative cradle-to-gate life cycle assessment of mid-rise office building construction alternatives: Laminated timber or reinforced concrete. *Buildings, 2*, 245–270.
26. Guo, H., Liu, Y., Meng, Y., Huang, H., Sun, C., & Shao, Y. (2017). A comparison of the energy saving and carbon reduction performance between reinforced concrete and cross-laminated timber structures in residential buildings in the severe cold region of China. *Sustainability, 9*(1426), 1–15.
27. Skullestada, J. L., Bohneb, R. A., & Lohneb, J. (2016). High-rise timber buildings as a climate change mitigation measure—a comparative LCA of structural system alternatives. *Energy Procedia, 96*, 112–123.

Chapter 10
One Floor at a Time: Cross-Laminating a Sustainable Future for Mass Timber in North America

Mona Azarbayjani and David Jacob Thaddeus

Abstract Cross-Laminated Timber (CLT) started in Germany and Austria in the early 1990s. As old-growth timber became difficult to find, glue-laminated timber (Glulam) and other engineered products became popular.

Since the advent of Mass Timber products around the world, architects, engineers, and manufacturers have leveraged CLT and other Mass Timber products for their advantages, including design flexibility, aesthetics, strength to weight ratio, and overall material performance. There is a revolution in the building sector that is working to elevate timber to the level of steel and concrete, mostly through taller and longer spanning structures.

In addition to many aesthetic and structural performance opportunities, and as a renewable resource with low embodied energy, CLT offers a low carbon footprint while also sequestering and serving as a carbon sink. Interest in CLT as a new engineered wood product in North America is still in the early stages of development and is widely and rapidly proliferating. Demand for CLT in the USA and Canada is driven by architects and engineers requiring sustainable timber-based building products and systems (Mohammad et al., Wood Design Focus 22:3–12, 2012). While the classic wood frame construction is a sound and economical option for smaller residential construction, CLT provides an opportunity for the North American wood industry to build both larger and taller structures in wood. However, development is delayed by building regulations that cautiously safeguard the public health, safety, and welfare whenever a novel building product is introduced, materiality notwithstanding. This chapter is intended to advance the knowledge of Cross Laminated

M. Azarbayjani, PhD (✉)
Associate Professor, University of North Carolina, Charlotte, NC, USA
e-mail: mazarbay@uncc.edu

D. J. Thaddeus
Professor, FAIA, NCARB, University of North Carolina, Charlotte, NC, USA
e-mail: thaddeus@uncc.edu

A. Sayigh (ed.), *The Importance of Wood and Timber in Sustainable Buildings*, Innovative Renewable Energy, https://doi.org/10.1007/978-3-030-71700-1_10

Timber technology. In addition, it will examine trends in mass timber construction in North America from the perspective of carbon footprint, structural performance, fire, and life safety. To that end, we are describing a legacy of excellence of Mass Timber construction in sustainability and tall building design and construction.

This chapter consists of:

Introduction:

- *Rings of time*.
- *Availability and abundance*: A naturally warm resource.
- *CLT, NLT, DLT, GLT, MPP, LVL*: Ingredients in the alphabet soup for the mass timber soul.
- *Organic paths of resistance*: Structural strength, directionality and grain.

Sustainability and Performance*:*

- *Time and time again*: A most renewable natural resource.
- *The trees they do grow high*: Innovation and the promise of technology in tall timber buildings … *and the leaves they do grow green*: Embodied energy, sustainability and green building.
- *One tree at a time:* Carbon footprint, carbon sequestration, and finding the forest through the trees (FFTT).

Case Studies
Conclusion

- *Cross laminated timber:* Leading the way into a sustainable future for mass timber.

References

Keywords Rings of time · Timber and time · Forest through the trees · Timber buildings · Mass timber construction · Cross-laminated timber · Timber reuse · Life cycle assessment of mass timber · Recycling and demolition · Bullitt Center · Brock Commons · T3 Minneapolis · John W Olver Design Building · Prefabrication

10.1 Introduction

Rings of Time

Highlights of the evolution of Wood in Construction in the USA.

The colonization of America by the British was politically motivated to extract resources from a rich colony while employing otherwise idle workers. In the New World was an endless supply of old growth trees. Many workers were sent to America to fell trees and ship them back to the British Empire from Jamestown and New England. By 1682, there were two dozen sawmills in Maine [1].

Eventually the lumber industry moved westward across the Great Plains and to the Pacific Northwest. Forests were devastated as the domestic housing industry grew and exports to European countries increased exponentially. This gave birth in the late nineteenth century to the Federal Reserve Act that protected large tracts of forests from logging. As a result of this act, loggers were required to make devastated forests productive again.

A pivotal ring in the growth of wood and timber construction in the USA was The Great Fires of Chicago in 1871 and then 3 years later in 1874. The 1871 fire spread over a four-square mile area and resulted in 300 deaths and left 100,000 people homeless. An estimated 17,500 buildings were destroyed [2]. Most of the structures that were destroyed had highly flammable roofs and were framed in wood. After the first fire, roof materials were replaced with terra cotta but the framing of the walls, floors, and roofs remained in wood. By the time the second great fire struck in 1874, there still were no fire protection requirements for buildings. Insurance policies were cancelled and were not renewed until changes to the fire and building codes were implemented.

As a result, Chicago and New York City banned the construction of wood structures within city limits. Exiled into the suburbs, wood frame construction enabled and greatly facilitated suburban sprawl.

Traditional heavy timber Post and Lintel construction that arrived to North America with the pilgrims was eventually replaced with Balloon framing around the middle of the nineteenth century. Balloon framing required a larger number of members but of smaller size in comparison to the fewer but larger timbers needed in Post and Lintel construction. In addition, the Post and Lintel system required a greater level of skill to craft the connection of members. The balloon frame remained the predominant residential wood framing system for nearly the next 100 years.

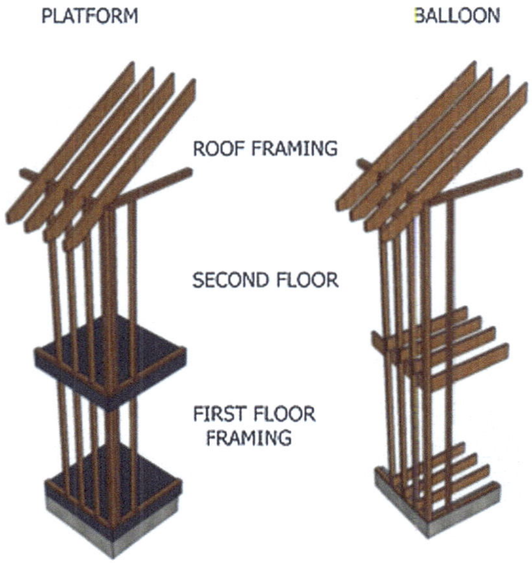

Platform versus Balloon framing

As aircraft and ship building activities ramped up during World War II, plywood, steel, and cement (concrete) were drafted into the fight against the Nazis Back on the home front, dimension lumber was left behind to substitute for the void that the war had left in the construction industry. As Rosie the Riveter became the star of the campaign to recruit women to work in the defense industry, so did lumber become the poster child for the residential construction industry. Soon after World War II, platform framing became the most prevalent framing technique [3]. With the availability of longer studs on the decline coupled with the difficulty of the acrobatics of assembly, platform framing has all but replaced balloon framing since the middle of the twentieth century.

Rosie the Riveter

As old-growth timber became difficult to find, Glue-Laminated timber (Glulam) and other engineered products became popular. The US Department of Agriculture Forest Products Lab in Madison, Wisconsin, commissioned the first glulam structure in 1934 [4]. The use of glulam was greatly amplified with the introduction of water-resistant adhesives in 1942 which laid to rest any concerns about glue degradation and possible delamination of individual pieces. With the introduction of glued finger joints, length of members was limited only by highway transportation regulations.

Christ the Light Cathedral. SOM. Oakland, California

Research on Cross-Laminated Timber (CLT) started at Graz University in Austria as the dissertation topic of Gerhard Schickhofer. His applied research and software development were presented in 1994 and earned him a PhD [5]. The first guidelines for CLT construction were refined and issued in Austria in 2002. In North America, Canada was quick to adopt CLT after it became very popular in Europe. After Canada imported the manufacturing technology from Europe, Oregon and Washington in the Pacific Northwest of the USA started designing and building projects with Canadian mass timber products.

Mass Timber construction at Wofford College, in South Carolina

No background on mass timber may be considered complete unless it addresses sustainable forest management. From the perspective of reforestation, conservation, and responsible forest management, voluntary efforts have resulted in agreements such as the Forest Stewardship Council (FSC) that emerged after the 1992 Earth Summit in Rio failed to produce any tangible results on deforestation [6]. More recently, the 2020 Davos Summit aimed to unite and promote reforestation efforts worldwide. At Davos, The World Economic Forum launched a global initiative to grow, restore, and conserve one trillion trees around the world. Reforestation efforts worldwide are emerging to protect forest resources from deforestation, forest fires, and to slow down climate change.

FSC

FORESTS FOR ALL FOREVER™

Forestry Stewardship Council

Worldwide timber technology and construction are currently enjoying an exhilarating renaissance with the introduction of Mass Timber products such as Cross-Laminated Timber (CLT), Nail-Laminated Timber (NLT), Dowel-Laminated Timber (DLT), Mass Plywood Panels (MPP), and others. CLT was first appeared in North America in 2011, which, along with other mass timber products, is the subject of this chapter [7].

Patterns of Growth: Direction, Strength, and the Ways of Nature

The hygroscopic nature of wood makes it able to hold indoor moisture while providing superior Indoor Air Quality (IAQ) and healthier interiors for human habitation. Unlike other construction materials that deliver hermetically sealed and insipid indoor environments, wood is biophilic, warm, and brings a touch of nature into the everyday.

Mass timber in general and Cross Laminated Timber in particular are a category of "engineered" yet natural wood products. It is engineered to regulate its properties and dimensions. To that end, CLT uses only kiln-dried (12% ± 3% Moisture Content) lumber that is finger-jointed to make longer individual pieces that are then joined together with structural adhesives [7]. The adhesive is engineered to resist a minimum threshold of heat while preventing the structural member from delaminating. The outer layers of a CLT panel may be graded for strength (machine stress-rated) or for visual appearance (fewer knots).

The Snow Family Outdoor Fitness and Wellness Center. Clemson University. South Carolina

Ingredients in the Alphabet Soup for the Mass Timber Soul

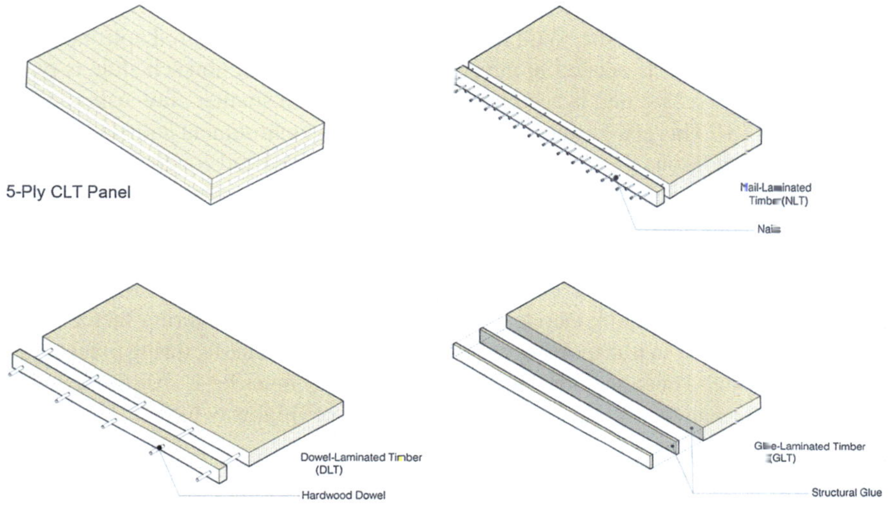

NLT, DLT, GLT, and other mass timber products are non-isotropic as they have greater strength in one direction compared to another. Unlike its mass timber cousins, CLT is theoretically more isotropic because it laminates timber in two perpendicular directions, alternating the grain in odd and even laminations. Whereas its cousins are one-way structural systems, CLT exhibits two-way structural action and behaves similar to a two-way concrete flat plate. This two-way action allows CLT to cantilever in two directions more easily at corners. In addition, for moderate spans, two-way behavior allows a CLT panel to be supported directly on a column without much concern for two-way punching shear.

Organic Paths of Resistance: Structural Strength, Directionality, and Grain

The greatest advantage of mass timber products is that they often employ damaged trees that would otherwise disintegrate and rot. Whether damaged by pine beetles or other destructive insects, the remains of some trees are too short or too small to be used in standard dimension lumber sawmills. Another distinction in the manufacture of mass timber compared to dimension lumber is that the former is able to use smaller diameter trees which cannot be used to produce dimension lumber. With finger jointing and quality control during manufacturing, shorter pieces of wood may be joined to make pieces that can be combined to produce longer mass timber members.

The prefabrication of mass timber components in a factory environment all but eliminates weather as a factor in construction delays. In fact, the factory environment assures quality, precision, and consistency of product and the timely delivery of mass timber members to the jobsite. With mass timber, more skill is required upfront at the fabrication stage than is needed in the erection stage. The erection schedule is typically shortened by 25% and is accomplished with a much smaller crew using much simpler tools [8]. This also results in a much safer work environment and less hazardous working conditions. In addition, when compared to steel and concrete, the lighter timber components require a smaller crane, which may be a great advantage in an urban setting. Noise pollution is also greatly reduced on a construction site and so is any on-site construction waste. Shorter construction schedules and precision milling that ultimately produce more energy-efficient envelopes make mass timber price competitive with steel and concrete construction. The obvious limiting factor is the restrictions involved in transporting the members from the manufacturing plant to the construction site. These include lane widths as panels must be 8′-10′ to fit in 12′ lanes; and less than 60′ in length to easily maneuver off-highway turns [7].

Image courtesy of Pixabay

Cross Laminated Timber panels have high dimensional stability that delivers in-plane and out-of-plane strength and stiffness. In resisting gravity loads, CLT panels are used as walls, floors, and roofs. In addition, in mid-rise construction, they may also be used as shear walls to resist lateral wind and seismic loads. The panels are durable and yet more ductile in a seismic event than unreinforced concrete, masonry, stone, CMU, and other brittle construction materials. The Seismic Response Modification factor, R, is currently being estimated at a value of 4 [9]. The creep behavior of CLT is a definite disadvantage as it deflects over time in ranges greater than glulam.

CLT and other mass timber products have high strength to weight ratio. This is a tremendous advantage as the resulting lighter structure will require less robust members to support it and also a smaller foundation which in the end is more sustainable. With the lighter structure also comes smaller values of base shear in a seismic event. Another sustainable feature of the spanning capabilities of CLT is that, in certain configurations, it may be able to eliminate perimeter beams. This in turn leads to taller apertures in exterior walls that will allow abundant natural daylight.

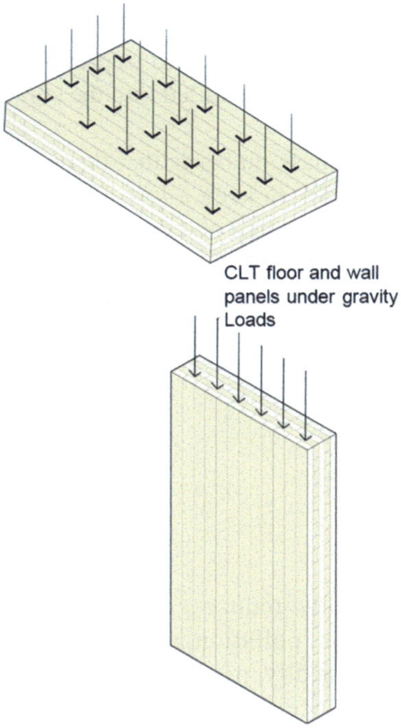

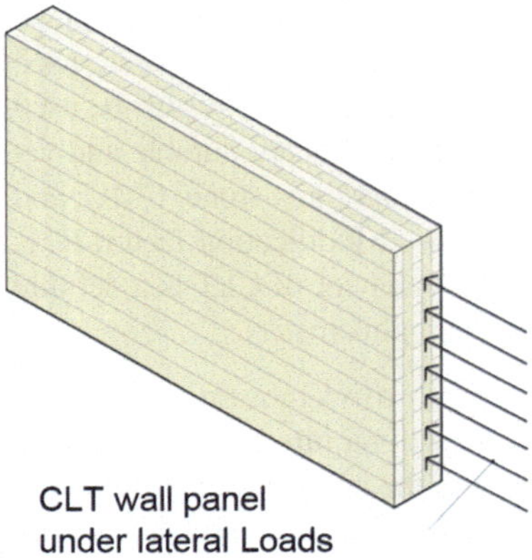

Mass timber floors and walls have some disadvantages that are easily remedied. These include vibration which can be reduced by the addition of a concrete topping slab on floor panels that will also help increase their fire resistance. A topping of 1″ also increases the Sound Transmission Coefficient (STC) of the panels [10]. Additional layers to sandwich mineral wool is a common practice to improve the acoustic qualities of a space. With DLT panels, grooves can be milled into the component members to improve the acoustics of each panel.

A significant but temporary local disadvantage of mass timber is that there is a very small number of manufacturing plants in the USA to keep a low embodied energy of the product. This will hopefully change very soon as 4 new certified plants have begun operations since October 2018 and 5 more are operational in the USA but do not yet hold a certification status [11]. With more supply and more built examples, building codes will have to catch up with the demand.

10.2 Sustainability and Performance

Introduction

Wood construction has the potential to substantially reduce greenhouse gas (GHG) emissions in the building sector, reduce waste, pollution, and costs associated with construction, while creating a more physically, psychologically, aesthetically pleasing, and healthy built environment. Since the dawn of time, wood has been used to build structures for human habitation, but after the Great Chicago Fire, wood structures came to be seen as unsafe and unstable when compared to the two materials that have since become the dominant construction materials worldwide, namely, concrete and steel. Residential buildings that are lower than three stories in the USA are economically constructed of dimension lumber. A novel way of using wood in a structural capacity is known as "mass timber" and has reinstated wood into a prominent position in the palette of construction materials. Not until recently, the construction of mid-rise and high-rise buildings consisted predominantly of steel and concrete. This is changing as mass timber is introduced into the construction market. Studies that take economic factors and structural strength into consideration find mass timber to be most suitable for mid-rise buildings [12].

When considering the full lifecycle impact of mass timber on carbon emissions, three factors need to be considered: **First,** beginning with forestry, greenhouse gas emissions are released in the supply chain. In logging, carbon in the soil is disturbed and released. Additionally, plant and wood waste are generated through the production process. This waste will eventually rot and release carbon. Emissions are also generated by the vehicles and equipment necessary to harvest lumber and transport it to the mill for production. Nonetheless, most conventional lifecycle analyses consider wood products to be carbon-neutral, when sourced from sustainably managed forests.

Second, there is an amount of carbon embedded in the timber itself, where it continues to be sequestered in wood members and could last anywhere from fifty to hundreds of years. The exact amount of CO_2 sequestered depends on the tree species, forestry practices, transportation costs, and a number of other factors. **Third**, and most significantly, substituting mass timber for concrete or steel avoids all embodied carbon involved in the extraction and production processes of these building materials. These processes are generally considered to be substantial when compared to mass timber. Cement manufacturing and the production of concrete are responsible for almost *8% of global GHG emissions*. The global iron and steel industry is responsible for another five percent. Approximately half a ton of CO_2 is emitted in the manufacture of a ton of concrete; while *2 tons of CO_2 are emitted* in the manufacture of a ton of steel. The embodied energy in the production of either concrete or steel holds the potential for great reductions in CO_2 emissions when substituted with viable mass timber products (Fig. 10.1).

Production

As structural elements, mass timber involves combining pieces of softwood together to form larger assemblies. Mass timber encompasses products of various sizes and functions, such as glue-laminated (glulam) beams, Laminated-Veneer Lumber (LVL), Nail-Laminated Timber (NLT), Dowel-Laminated Timber (DLT) and Glue-Laminated Timber (GLT). The most common and most familiar form of mass timber, Cross-Laminated Timber (CLT), has "opened the door" to a wide range of architectural possibilities. In general, mass timber products are manufactured from sustainably grown trees and responsibly managed forests (Fig. 10.2). These forests are often considered repositories for carbon. Through the photosynthesis process, carbon dioxide is converted into sugar and other organic molecules which make up living organisms (biomass). Biomass is defined as the total mass of organisms in a given area or volume. All forests are considered carbon repositories as plants and trees

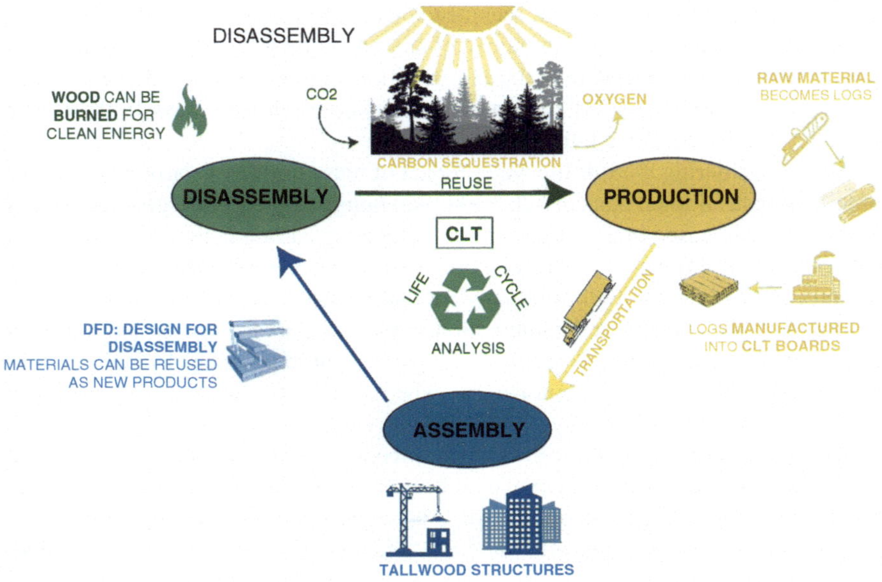

Fig. 10.1 Life cycle assessment of mass timber

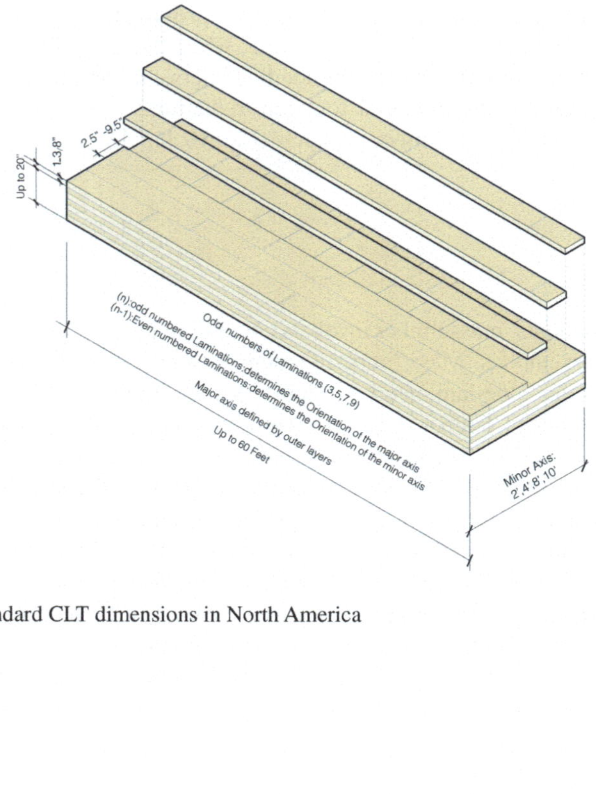

Fig. 10.2 Standard CLT dimensions in North America

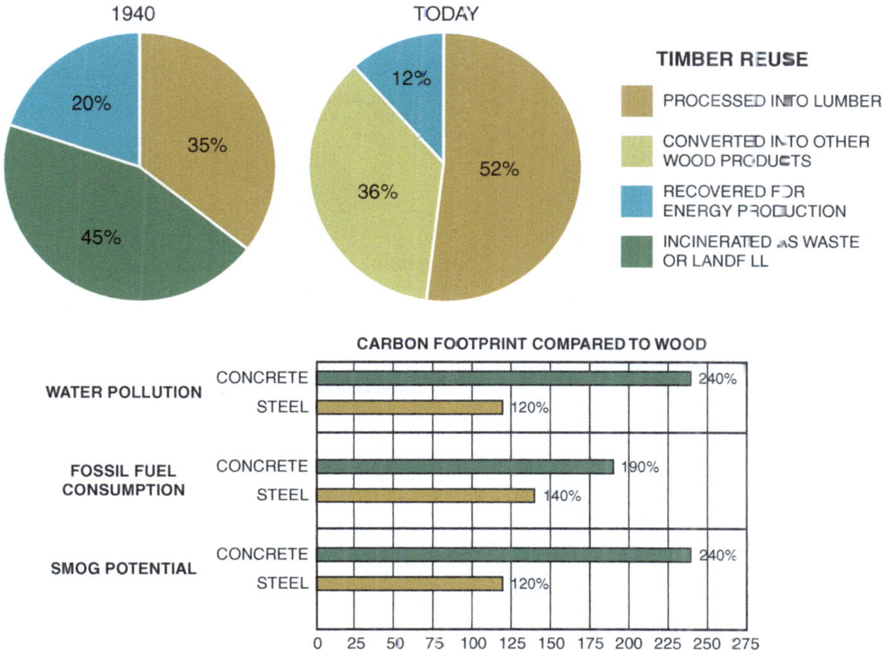

Fig. 10.3 Timber production data comparison to steel and concrete

sequester carbon and release oxygen through the process of photosynthesis. Water and fossil fuel demand for the manufacture of timber products is far lower compared to the production of concrete or steel. In addition, this results in less greenhouse gas (GHG) emissions and a smaller carbon footprint of a building (Fig. 10.3).

In addition to the renewable nature of wood itself and its low-impact method of production, the abundance of trees in North America reduces the energy consumed by importing and shipping the product from distant locations [12]. The wood utilized in the production of CLT is usually fast-growing spruce softwood from conifer trees in the Northwest USA and pine-fir in the Southeast. These are mostly small section lamellas [13]. Sustainably managed forests ensure that reforestation efforts stay well ahead of deforestation activities. Future predictions show a twofold increase in the demand for mass timber products by 2050. New forestry resources will need to be developed if this enthusiasm for mass timber is to be met [14]. As deforestation is one of the most significant factors in climate change, sustainable forest management is a fundamental focus that needs to be considered as the rate of harvesting trees accelerates [15]. The fabrication of CLT products does result in minimal wood waste. However, this additional wood can be recycled or utilized in other wood-based projects such as mulch, or firewood ensuring that no wood goes to waste [16]. Other less usable waste may be converted to energy or compressed

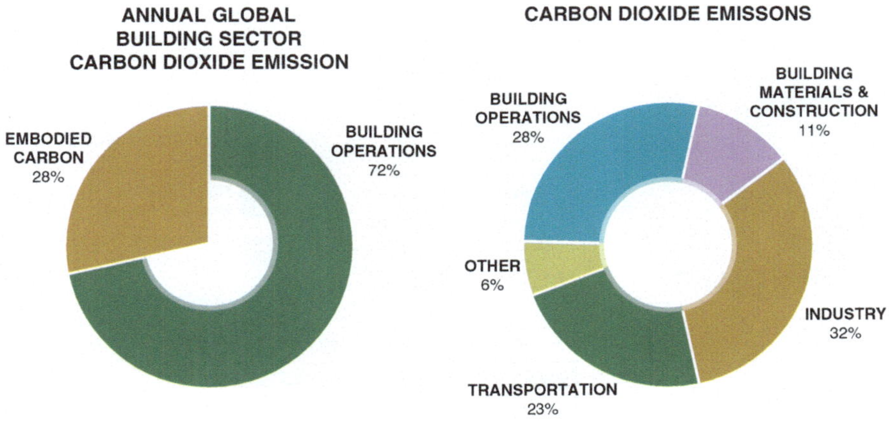

Fig. 10.4 Carbon footprint by building sector data

into shippable pallets for use as an energy source similar to coal [14]. Because several prefabricated components can be installed in sequence and with relatively little labor, they can be shipped to the construction site on a just-in-time basis, thus avoiding massive on-site inventory and minimizing on-site disruption. Even tall towers have the ability to be constructed within weeks, with low labor costs.

Recent reports and publications about mass timber buildings with new structural systems challenge concrete and steel as the only sustainable material option for tall building design [15]. Mass timber promises to revolutionize the construction industry with endless possibilities for genuinely sustainable development. CLT is becoming very popular for its architectural and aesthetic qualities and for its potential to help decarbonize the building sector. Approximately 11% of global greenhouse gas emissions comes from building material production and construction activities, while another 28% comes from building operation and management. Buildings generate nearly 40% of the annual global GHG emissions. Within this percentage, nearly 72% of carbon dioxide emissions are produced during building operations, and only 28% from embodied carbon in existing structures (Fig. 10.4). This data confirms that the use of cleaner materials in the construction of buildings could have the largest impact on reducing carbon emissions in the built environment. Mass timber is proving to be the most efficient material choice to meet this goal.

Comparisons of mid-rise and high-rise CLT commercial buildings to those with similar programmatic and functional requirements, but alternatively constructed of reinforced concrete, have been conducted on a large scale. One comparison concluded that a CLT building represents a "26.5% reduction in global warming potential" [17]. Additional research has compared the embodied carbon of CLT buildings with that of reinforced concrete buildings, which is shown to be 655 metric tons of carbon dioxide equivalent. This "is 1006 metric tons of carbon dioxide equivalent lower than the RC frame equivalent" [13]. A general rule for this comparison states that one cubic meter (424 Board Feet—BF) of wood sequesters roughly one metric

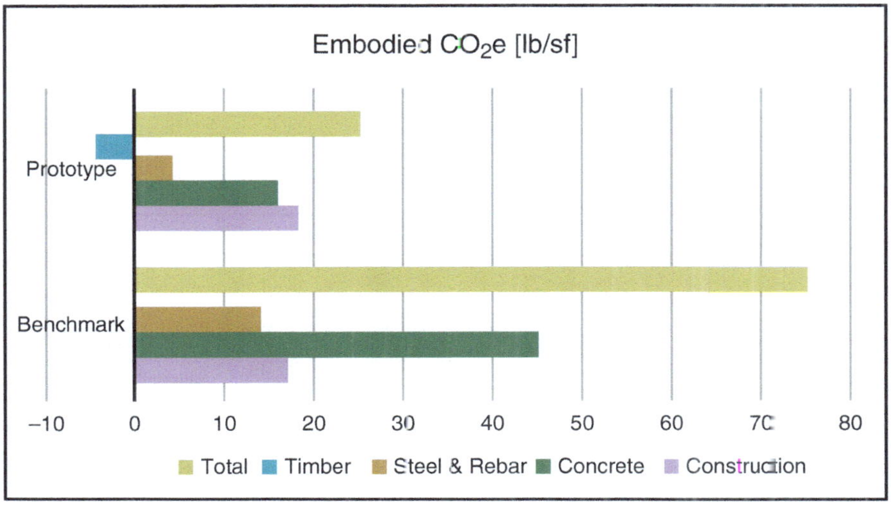

Fig. 10.5 Embodied carbon comparison

ton (1.1 US tons) of CO_2. Although this is generally an accurate benchmark value, this equivalence can be adjusted up or down to reflect forestry, transportation, milling, construction, and disposal practices [13].

Construction and Operation

In addition to energy consumption and CO_2 emissions of buildings during the operational phase, it is necessary to consider the carbon emissions during the construction phase. This primarily includes material extraction, processing, manufacture, and transportation. These factors contribute significantly and quite adversely to the energy profile of a building. The energy consumption at the inception stage of any material is a leading indicator of its rank on the sustainability scale (Fig. 10.5). Considering the construction phase of a project, the energy consumption of concrete frames has been shown to be 30% greater than that for timber frames under various scenario frameworks [18]. The results of a lifecycle comparison between Reinforced Concrete (RC) and CLT buildings show that the energy consumption of CLT buildings is 9.9% lower than RC buildings with 13.2% higher carbon emissions [18]. Other studies show 50% less CO_2 emissions in CLT buildings compared to RC buildings with regard to material extraction, transportation, site work, and end-of-life disposal [19]. According to the "Carbon, Fossil Fuel, and Biodiversity Mitigation with Wood and Forests" study, 14–31% decrease in global CO_2 emissions could be achieved by using wood as a primary construction material [12].

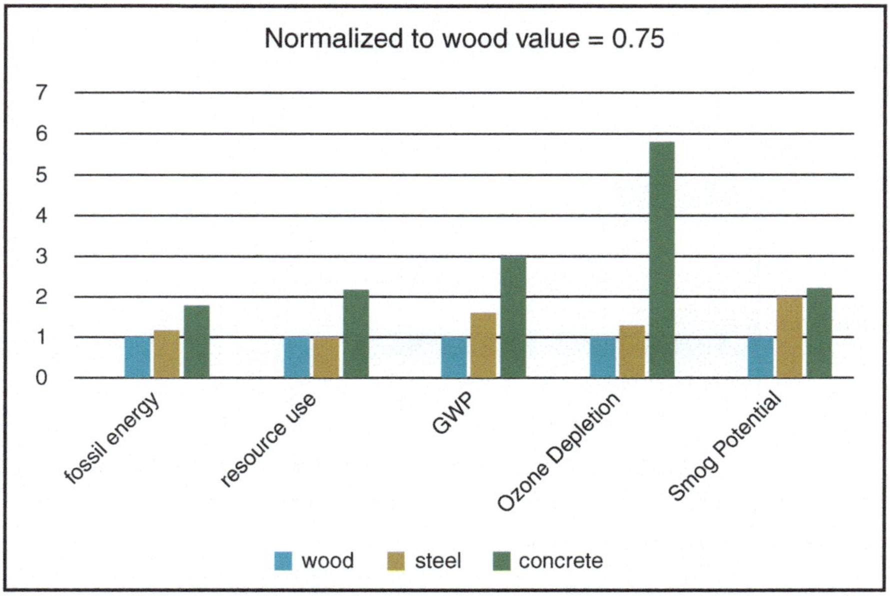

Fig. 10.6 Construction material data comparatively

CLT construction involves lower labor costs and less on-site construction waste than alternate materials (Fig. 10.7). Rather than materials being ordered in mass quantities, cut to size on site, and assembled, as with conventional construction, much of the labor and fabrication of CLT buildings is completed before the materials are sent to the site. It is often fabricated using "computer numerical control" (CNC) machines to allow precision cuts. A CLT wall can be fabricated exactly to specifications provided by architects and engineers. Door and window openings can be created in exact locations and take into account provisions for electrical and plumbing installations. This process virtually eliminates material waste and ensures that all openings are predetermined and accounted for.

Although studied in depth, there are contradictory opinions in the academic community regarding the correlation of carbon storage and timber efficiency in CLT buildings. Carbon storage is defined as the embodied biogenic carbon in the material over its lifetime, and is discharged to nature at the end-of-life stage. Based on the research, the Global Warming Potential (GWP) is different in timber-based buildings. CLT materials have been shown to contain a lower GWP in comparison to steel and concrete frame construction (Fig. 10.6) [20]. Since the thermal mass of the CLT is lower than that of concrete, the amount of stored heat is lower by about 2.5 times. This emphasizes the importance of mechanical and ventilation considerations, specifically in office buildings. Some argue that these mechanical ventilation requirements will cancel out the negative footprint of the building over its lifetime. However, additional mass that is necessary for increasing the heat storage capacity of CLT may also have a negative influence on the initial carbon sequestration credit of this material [19]. Nonetheless, the low thermal conductivity of CLT has the potential to make a completed building more airtight by decreasing the effects of thermal bridging.

Recycling and Demolition

Compared to alternate materials, CLT can be recycled and disassembled relatively easily and efficiently. In order to recycle and reuse full CLT panels and boards, each timber element must be separated individually through the process of removing the adhesive which joins the elements together. Currently, there are no proven procedures to complete this task at the scale of a building. Therefore, it is much more common and effective to utilize the CLT elements as waste materials [21]. The process of incorporating CLT panels into waste material involves the separating of CLT elements into smaller pieces, which are then used to create wood composite panels. These wood composite waste panels can range from chipboards (particle board) to wood bonded panels, and even wood fiber insulation panels. Each of these recycled wood elements can be utilized structurally in new mass timber projects. Particle board has been proven to be effective when positioned on the surface of walls and roofs, and is commonly used as floor decking. Wood bonded panels and wood fiber insulation panels have demonstrated excellent thermal insulation and are commonly used in residential buildings. Due to the transformation of CLT into other wood products, one study estimates that it is possible to consider CLT 100% sustainable [21].

Aside from the wood boards, excess wood waste can be converted into a multitude of different elements, even pulp and paper. In New Zealand, some emerging companies have turned excess waste from this process into mulch that can then be used for composting or fuel for generating heat. From this process, over three and half million cubic meters of wood chips and sawdust are generated into energy every year in New Zealand processing plants [22]. Although methods of recycling CLT waste into alternative materials are becoming more common, landfills are still the most common method of disposing of municipal waste. Of this total municipal solid waste, 69.5% is biomass, and 5.8% of this biomass is timber [22]. While this is a relatively low percentage, CLT is available for a number of recycling options and has the potential to be 100% reused. Through the education of recyclable options for CLT, the percentage of timber in landfills has the potential to reach 0%.

A recent study conducted in China found that the energy required for the demolition of a building can be assumed to be about 90% of the energy that was required during the erection phase of the structure [23]. According to the softwood lumber industry, "a mass timber project is approximately 25% faster to construct than a

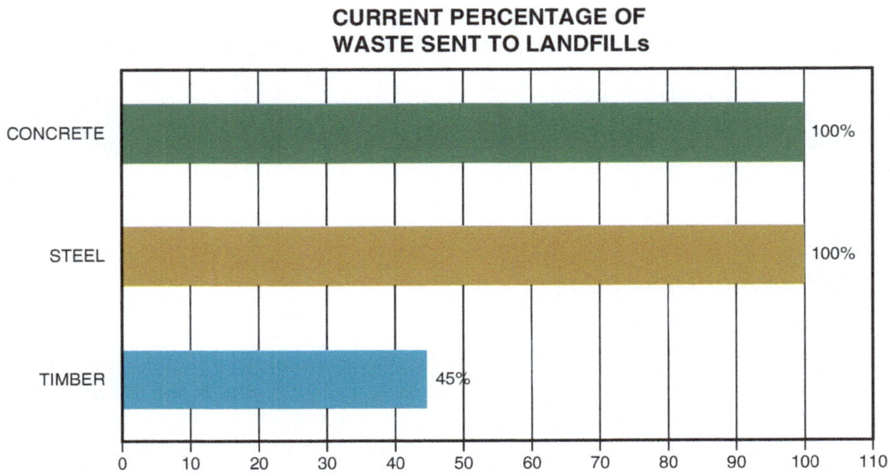

Fig. 10.7 Current percentage of waste sent to landfills

similar project in concrete… [and] offers 90% less construction traffic" [8]. Given that mass timber buildings utilize prefabrication and the erection phase is much quicker compared to buildings constructed of alternative materials, it is safe to assume that the demolition of mass timber buildings also requires less energy. In the same study, buildings in the Chinese cities of Harbin and Xi'an were studied and data determined that 100% of all the concrete and steel materials after demolition were sent directly to landfills. Alternatively, CLT buildings in these two cities showed a recycle rate of 55%, with an assumed 45% used for biomass energy (Fig. 10.7) [23]. Although far from the 100% recycle rate discussed above, it is clear that even if all measures to reuse CLT waste are not implemented, the rate in which timber is recycled compared to materials such as steel and concrete is significantly higher. When recycled, timber elements continue to reduce the carbon emissions of the new projects in which they are employed while saving time and money on new materials.

10.3 Case Studies

This section will examine trends in mass timber construction in North America from the perspective of carbon footprint, structural performance, fire, and construction. To that end, we are describing a legacy of excellence of Mass Timber construction in sustainability and tall buildings. Case studies that are described are as follows:

1. The Bullitt Center, Seattle, WA (2013). Greenest building in the world.
2. Brock Commons Tallwood House, University of British Columbia, Vancouver, Canada(2013). Currently tallest wood structure in North America.
3. Hines T3-Minneapolis (LEED Gold). Recognizing the demand for sustainable commercialdevelopment, T3 (Timber|Transit|Technology) is paving the way in most innovative ways.
4. John W. Olver Design Building at the University of Massachusetts Amherst (LEED Gold) and AIA COTE Top Ten award 2020.

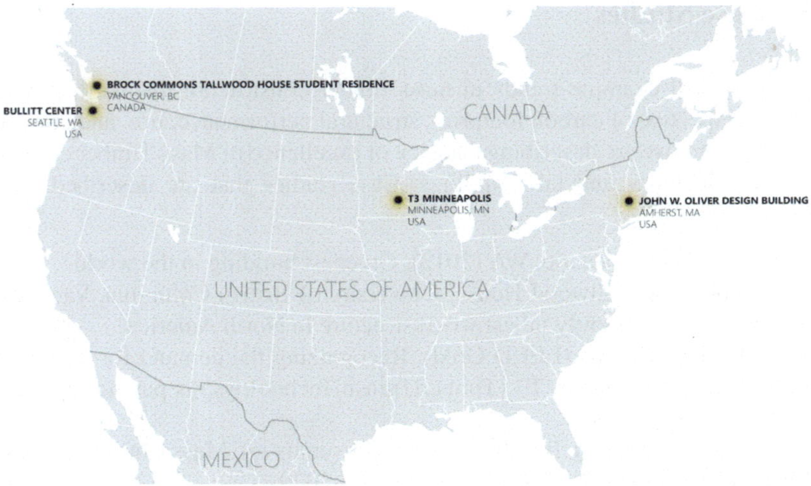
BROCK COMMONS TALLWOOD HOUSE STUDENT RESIDENCE
VANCOUVER, BC
CANADA
BULLITT CENTER
SEATTLE, WA
USA
CANADA
T3 MINNEAPOLIS
MINNEAPOLIS, MN
USA
JOHN W. OLIVER DESIGN BUILDING
AMHERST, MA
USA
UNITED STATES OF AMERICA
MEXICO

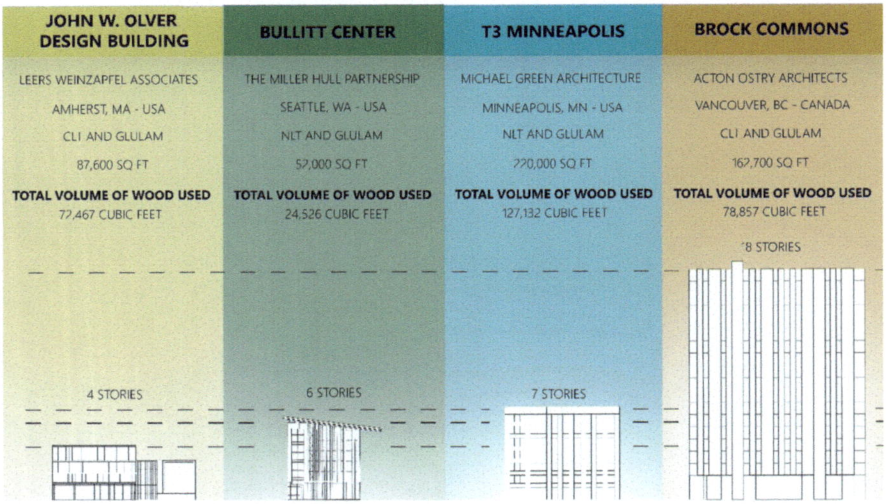
JOHN W. OLVER DESIGN BUILDING
LEERS WEINZAPFEL ASSOCIATES
AMHERST, MA - USA
CLT AND GLULAM
87,600 SQ FT
TOTAL VOLUME OF WOOD USED
72,467 CUBIC FEET
4 STORIES
BULLITT CENTER
THE MILLER HULL PARTNERSHIP
SEATTLE, WA - USA
NLT AND GLULAM
52,000 SQ FT
TOTAL VOLUME OF WOOD USED
24,526 CUBIC FEET
6 STORIES
T3 MINNEAPOLIS
MICHAEL GREEN ARCHITECTURE
MINNEAPOLIS, MN - USA
NLT AND GLULAM
220,000 SQ FT
TOTAL VOLUME OF WOOD USED
127,132 CUBIC FEET
7 STORIES
BROCK COMMONS
ACTON OSTRY ARCHITECTS
VANCOUVER, BC - CANADA
CLT AND GLULAM
162,700 SQ FT
TOTAL VOLUME OF WOOD USED
78,857 CUBIC FEET
18 STORIES

Bullitt Center, Seattle, Washington

Architect: The Miller Hull Partnership, LLP (Fig. 10.8).

Fig. 10.8 The Bullitt Center officially opened on Earth Day, April 22, 2013

Introduction

Often described as the "World's Greenest Building" or the "Greenest Commercial Building in the World," it is no surprise that the Bullitt Center, located in Seattle, Washington, does not only demonstrate sustainable design, but sets the bar for what a net-zero-energy building can accomplish. Designed by The Miller Hull Partnership LLP, the structure serves multiple purposes, acting as an educational facility on the two lower levels, and providing private commercial space on the floors above. Successfully meeting all the criteria of the Living Building Challenge (LBC), this structure satisfies difficult material procurement requirements, ones that are specifically related to the timber construction used on the upper four floors of the building. The decision to use timber as the primary structural element is the fundamental design choice that is predicted to enable the Bullitt Center to last 250 years after its inauguration [24]. Nail-Laminated Timber (NLT) was the material selected for the floor and roof deck, while the columns and beams were constructed of glulam, and wall panels were fabricated from plywood. Despite the total project cost of over \$30 million (approximately \$577 per sq. ft), a cost almost twice that of a comparable office building in the region, the sustainable design decisions made in the Bullitt Center will ultimately save its tenants money in the future while providing net zero energy on an annual basis [25].

Sustainable Features: Timber as a Structural Element

Before diving into the many innovative structural aspects of the Bullitt Center's renowned timber structure, it seems necessary first to discuss the materiality of the two levels of the building closest to the ground. The first two floors consist of reinforced concrete (manufactured less than 300 miles away in order to conform to the LBC guidelines) chosen for its thermal properties, despite the carbon footprint associated with concrete (Fig. 10.9). This technique is very common in large-scale timber structures as it allows different occupancies to exist in the same structure. According to the International Building Code (IBC), larger floor areas and taller heights are permitted if timber construction is built over a concrete podium. Employing concrete as a foundation helps secure structural stability and safety in this seismic zone. With building codes limiting the height of structures to conform with fire engine ladder reach, building with timber over a concrete podium provided adequate fire separation between occupancies. This permitted the use of timber structures for upper levels with a total height reaching four stories for unprotected construction types and five stories for construction types that have noncombustible exterior walls. Other than in the podium, concrete topping is used on upper levels of the building to facilitate natural ventilation and night flushing strategies. In addition, the concrete topping helps with sound insulation and vibration control. Apart from these concrete components, timber is the primary construction material in the remaining structure.

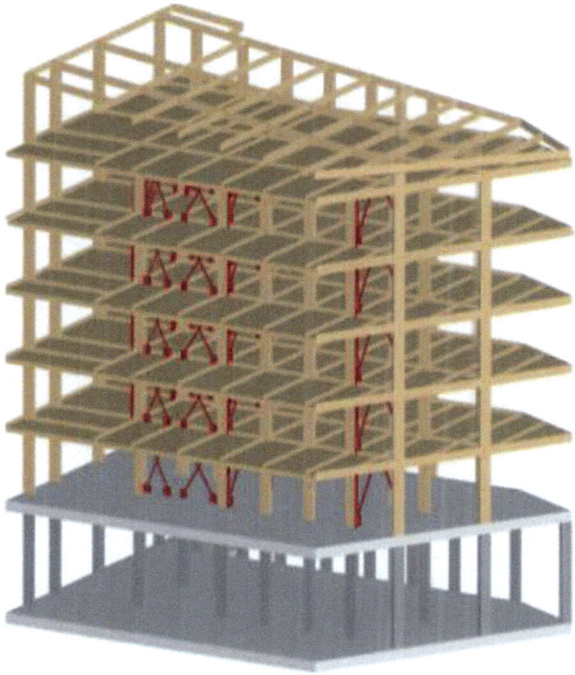

Fig. 10.9 The Bullitt Center. Schematic Section Axonometric

In contrast to the podium at the base of the building, glulam beams and columns became the logical choice over concrete due to the sustainable nature of wood. Manufactured in Vancouver, WA, by Calvert Glulams, each glulam beam and column was fabricated from smaller planks sourced from FSC (Forest Stewardship Council) certified forests. To meet the stringent standards of the Living Building Challenge, all wood harvested for the project was required to come from within a 600-mile radius of the site [25]. For this project the timber came from Douglas fir trees, a species local to the Pacific Northwest of the USA. Such a standard is logical in this case as timber construction is by far a more sustainable choice than steel. This holds true as long as the timber is not harvested from distant forests and sent to the site as carbon emissions involved in the transportation would only counteract the sustainable decision to use timber. In preparation for the vertical shrinkage of wood over time, each glulam joint required custom-designed steel connectors (Fig. 10.10). These steel tubes work as inserts between the timber columns to stop radial shrinkage in the beams from affecting the columns, and ultimately any cumulative shrinkage in the structure as a whole. The connection is designed to introduce enough bearing by overlapping the beams on the columns to survive a fire that would otherwise compromise the steel and cause the sudden and catastrophic collapse of all members framing into the connection. The insightful combination of glulam timber

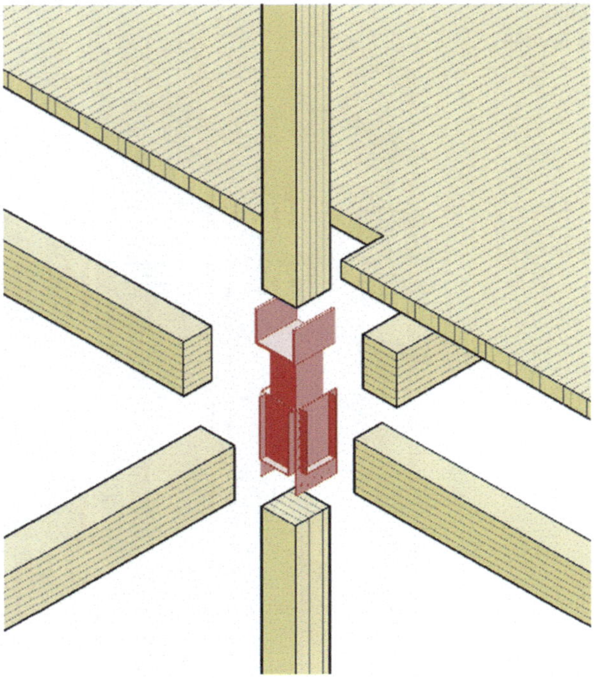

Fig. 10.10 The Bullitt Center. Assembly detail

and steel connectors prevents possible long-term issues and is the primary structural element of the Bullitt Center that has given it the potential to last up to 250 years into the future.

The selection of multiple structural systems and materials for this building exemplifies each for its advantages and strengths. Concrete is used in the basement and the foundation as they are in contact with earth. The high ductility that the steel bracing members manifest is employed to combat the appreciable lateral seismic forces in Seattle. The seismic base shear is greatly reduced with the use of timber since it is significantly lighter than both concrete and steel and thus produces smaller structural members and smaller foundations to support them. The mass timber structure is responsible for supporting the gravity live and dead loads (Fig. 10.11).

In conformance with the Living Building Challenge, on-site waste production was minimized and all concrete placed on the site was free of workability admixtures that are red-listed for containing toxins. Likewise, only water-based sealants were used and interior materials were left unfinished to avoid possible toxins. In essence, The Living Building Challenge requires that materials with the lowest embodied energy be used and materials be sourced from as near locations as possible and be free of red-listed materials such as VOCs, PCBs, and other toxic materials.

Fig. 10.11 View of custom-fabricated steel inverted chevron bracing and connectors joining glulam beams and columns and providing lateral support for the structure against seismic loads. (Image Credit: Letao Tao)

In addition to glulam columns and beams, Nail Laminated Timber (NLT) was used on the floor and roof decks. NLT was chosen primarily because its application allowed an extra two feet of ceiling height on each floor, increasing the height from the bottom of the beam to the top of the deck. This material choice aided in maximizing the interior ceiling heights, without increasing the overall building height. This aligns with one of the primary goals of the building, namely, to increase daylight penetration into the interior of the building. Extended exposure to sunlight, which is amplified by the use of NLT, is an important characteristic of the Bullitt Center. This in addition to the multifaceted skin of the building plays an important role in the heating and cooling of the structure.

The triple pane glazing skin system used on the exterior of the Bullitt Center is fabricated in multiple layers, each playing a key role in the heating and shading of the structure (Fig. 10.12). The high-performance envelope is expected to last up to 50 years [24]. The outermost layer of the skin involves functional stainless steel shades that are designed to scatter direct rays in the summer when cooling is necessary and maximize natural daylight in the winter when heating may be needed. Interior blinds allow the users to manually control the light exposure entering the building. This not only affects the amount of glare and light exposure in the building but can also be used to heat each individual space by allowing the maximum amount of sunlight in or cool down a space by limiting exposure to direct sun. The windows themselves are operable and can be opened to allow for individualized thermal

Fig. 10.12 Image of the multilayered envelope system on the exterior of the Bullitt Center. (Image Credit: Letao Tao)

comfort specific to each room. Catering to the inhabitant's needs during each of Seattle's contrasting seasons proved to be a main source of the design intent that led to the decision to use an operable building skin.

Aside from the presence of exposed timber as an interior finish, other types of wood employed in the Bullitt Center include ½" and ¾" plywood. ¾" plywood was used on the roof for lateral shear [25] which helped provide stability for the installation of the 575 solar panels mounted on the roof, covering an area of 14,000 sq. ft. [26]. ½" plywood was applied as backing panels on select walls and also used as part of the floor assemblies directly above the structural NLT flexible diaphragm. The plywood flooring was additionally topped with a noise barrier and an insulation mat applied directly below the thin concrete topping slabs mentioned above.

Case Study Conclusion

As the largest and first commercial building to achieve the Living Building Challenge Certification (LBC), it is no surprise that this ambitiously and sustainably designed structure is renowned as one of the greenest buildings in the world. Since its completion in 2013, the Bullitt Center has received recognition from 18 different committees regarding the sustainable choices implemented throughout the design and construction phases of the project [27]. The Miller Hull Partnership's decision to be bold and lead by example is one that has benefited the environment and illustrated the possibilities of a timber constructed building in the twenty-first century. The 24,526 cubic feet of timber used in the construction of the Bullitt Center can be regrown in the US and Canadian forests in only 2 min, a small price to pay for a building with a projected lifespan of 250 years. In addition, choosing timber instead of concrete or steel has provided a potential carbon benefit of 1703 metric tons of CO_2 and avoided greenhouse emissions by an amount of 1158 metric tons of CO_2. This is equivalent to the amount of carbon used to operate an average house for 145 years or the amount produced by 325 cars on the road during a year [25].

Carbon Footprint Graphic

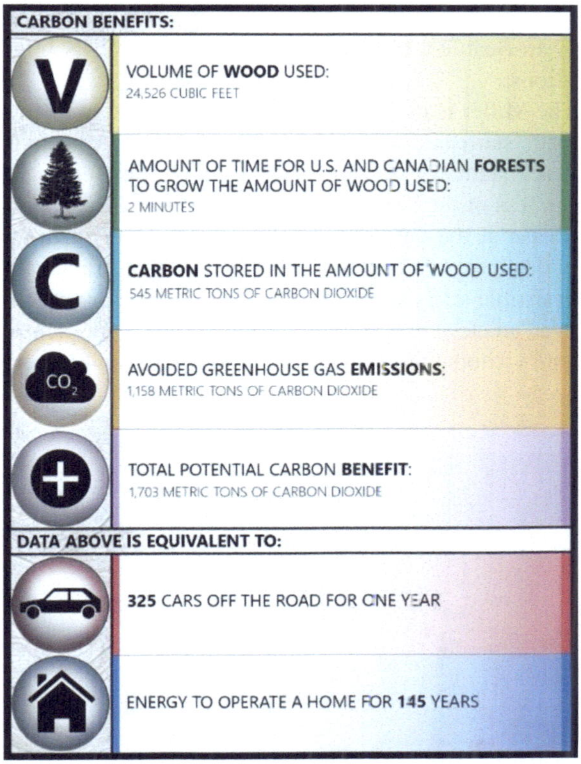

Additional Information: Sustainability Features: Energy

When examining the exterior of the Bullitt Center, it is virtually impossible to overlook the large overhanging roof system. Expanding past the base of the building, the roof system is home to 575 solar panels, covering 14,000 sf of space (Fig. 10.13). Although predicted to generate 257,770 kWhrs per year, the first year showed that the array of solar panels actually generated 251,885 kWhrs, 2.3% less than what was predicted [28]. However, the building used 41.7% of the energy that was predicted in the energy model, ultimately producing a surplus of 114,085 kWhrs of electricity. The diagram below graphically displays the actual versus predicted energy use and production from the Bullitt Center's first year in use. In essence, the Bullitt Center is proving to be self-sufficient and producing more energy than is currently being consumed and furthermore, the active solar control photovoltaics are expected to remain efficient for up to 25 years. When energy production is high during the summer months, the surplus energy is stored in Seattle's electrical grid and then retrieved in the winter months when production is low. This method is extremely effective, especially in Seattle's tropical oceanic climate and its infamously gloomy weather conditions.

In this six-story building, The Miller Hull Partnership used active design tactics to promote the use of the main staircase, also dubbed the "irresistible stairs." Choosing to walk up and down a staircase instead of using an elevator not only promotes human health but uses less of the harvested energy to power the elevator. The position of the main staircase at the entrance of the building along with an abundance of glass walls provides amazing views of the city, and that graciously invites occupants to use the irresistible stairs as the most inviting choice for vertical circulation (Fig. 10.14). However, an elevator is required in a large structure such as this, whether it be for tenants and visitors with disabilities, or for those transporting

Fig. 10.13 Bullitt Center Roof System from below. (Image Credit: Bruce Englehardt/CC-BY-SA)

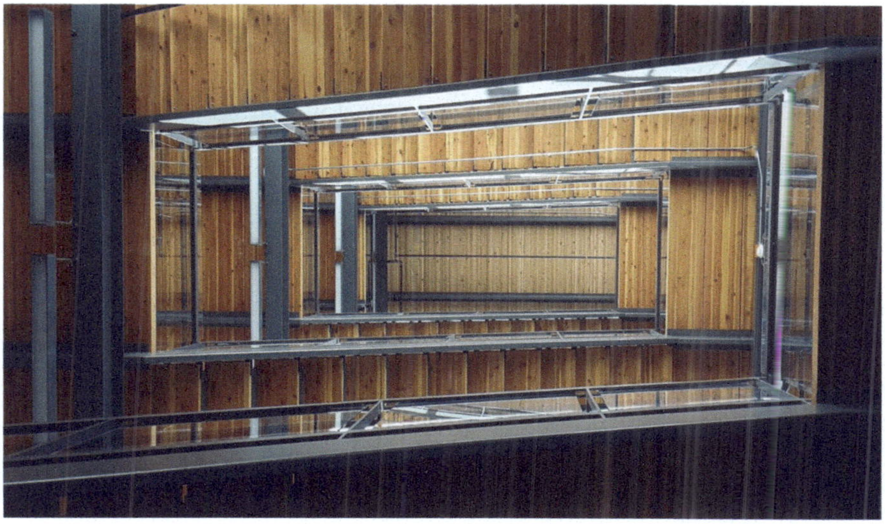

Fig. 10.14 View from inside of the "Irresistible Stair" looking up. (Image Credit: Brian Deck)

large objects. In an effort to create an elevator that uses as little energy as possible, the elevator in the Bullitt Center converts kinetic energy from braking into useable energy. This technique is said to make the elevator about 60% more efficient than the average elevator [29]. Independent of the mode of vertical circulation chosen, it is clear that sustainable design was at the forefront of the decision-making process.

Sustainability Features: Water

It should come as no surprise that the water usage in the Bullitt Center has also been carefully examined and designed with deliberate sustainable intentions. Just like the Douglas fir forests that provided the timber for the structure, the Bullitt Center returns water to the soil and atmosphere through rainwater harvesting. Along with the solar panels, the roof system is also designed to capture rainwater and store it in a 56,000-gallon tank below the building [30]. Large particles in the water are then filtered out and micron filters and UV light are used to treat any impurities before being used throughout the structure as greywater. It may come as a surprise that this recycled water is not used in flushing the plumbing fixtures in the Bullitt Center. However, this is due to the fact that the toilets in this structure are waterless and designed for composting. This is the world's only six-story composting toilet system. Human waste is sent directly to composters beneath the structure and is eventually taken offsite to be used as fertilizer.

A complex greywater system is used to ensure that the maximum possible amount of water is recycled back into the environment after its use in the Bullitt

Center. Any water that is used in the building is first stored in a 500-gallon greywater tank located in the lower-level basement [31]. The water is then sent to a constructed "wetland" (green roof) located on the third floor of the building that contains multiple horsetail plants, *equietum*, which soak up nutrients that may be harmful to the surrounding ecosystems. After multiple cycles, if water has not evaporated, it is released in the bio-swales along the western edge of the site and returned to the earth as groundwater.

Sustainability Features: Heating And Air

Keeping in accordance with the other sustainability features of this structure, rooms throughout the Bullitt Center are heated and cooled through a system contained below the building itself. Located 400 ft. beneath the structure, 26 geothermal wells contain a mixture of glycol and water contained at a constant base temperature of 53 °F [32]. This water and glycol mixture is then passed through a heat pump system which increases the temperature of the mixture to 95 °F. The heated water is then circulated through tubes that are imbedded in the concrete topping slabs located on each level of the building, naturally heating each level through the floor.

In addition to floor heating, ventilation plays an important role in the heating and cooling systems used in the Bullitt Center. Both functional exterior shades and interior blinds allow the users to manually control the light exposure entering the space (Fig. 10.15). This not only affects the amount of glare and light exposure in the

Fig. 10.15 Exterior shades. (Image Credit: Letao Tao)

building, but it can also be used to heat each individual space by allowing in the maximum amount of sunlight, or cool down a space by limiting exposure to direct solar gain. The windows themselves are operable and can be opened or closed to allow for individualized thermal comfort in each room of the structure.

In addition to the operable skin of the structure, a heat recovery ventilation system is used to provide the building with warm fresh air. Both fresh outdoor air and stale indoor air are cycled within this system before being released throughout the building. Exterior air is filtered to remove possible toxic particles or smoke, while stale warm interior air from bathrooms and kitchens is purified to separate clean air from CO_2 or moisture. After the filtration process, warm fresh air is sent in the building while the stale air and CO_2 are expelled to the outside. This process transfers approximately 65% of the heat from the warm exhaust air into new incoming air [33].

Brock Commons Tallwood House Student Residence
Vancouver, British Columbia. Canada. Architect: Acton Ostry Architects Inc. (Fig. 10.16).

Introduction

Holding the title of the tallest mass timber structure in the world at the time of its completion, Brock Commons Tallwood House Student Residence was completed in 2017 and reaches a height of 174 ft. This 18-story hybrid building was designed by Acton Ostry Architects, Inc. in collaboration with Hermann Kaufmann Architeckten as the tallwood timber advisor [34]. Located in Vancouver, Canada, and functioning as a mixed-use residential dormitory for the University of British Columbia, Brock Commons required an immense amount of unique problem-solving, case-specific building-permitting, and in-depth code research. At the time of its construction, the prevailing British Columbia Building Code (BCBC 2012) limited the height of timber construction to six stories. As a result, provisions from the not-yet-adopted 2015 National Building Code of Canada (NBCC) were required in addition to approval of performance aspects that involved comprehensive scientific data that validated the building's safety [35]. The University of British Columbia resorted to a Site-Specific Regulation (SSR) and appointed several peer review panels to confirm the validity of the structural and fire-resistance proposals for this building. In compliance with stringent structural and safety requirements, additional project-specific aspects of the building had to be considered and studied. Despite the additional work required to design and construct a timber building of this size, the Brock Commons Tallwood

Fig. 10.16 Brock Commons Tallwood House Residence Hall. (Image Credit: naturallywood.com. Photographer: Brudder)

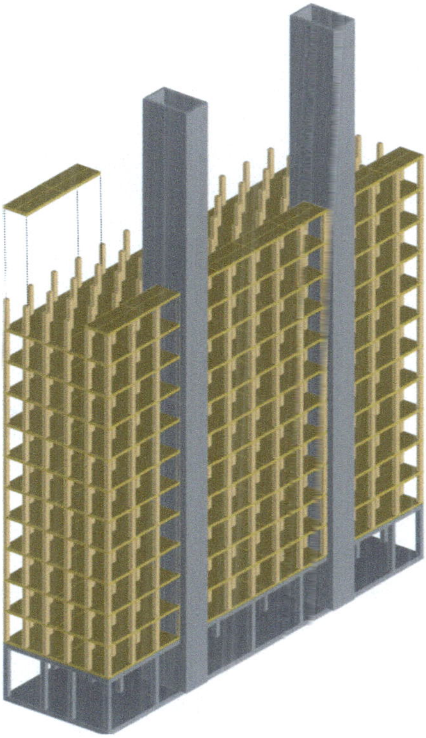

Fig. 10.17 Brock Commons Tallwood House. (Schematic Axonometric)

House Student Residence reached its goal of LEED Gold and proved that a high-rise timber building was not only feasible, but optimal for sustainable design intent. At the time of this writing, a 25-story mass timber building named The Ascent in Milwaukee, Wisconsin, is to break ground in the fall of 2020 and will eclipse Brock Commons as the tallest mass timber structure in the world (Fig. 10.17).

Prefabrication and Timber Construction

Not unlike most large-scale mass timber buildings, prefabrication is an essential stage in the construction of the structural elements. For this building, a two-story mock-up was constructed from a Virtual Design and Construction (VDC) modeler that was highly detailed and that was used for the fabrication of all mass timber members. The glue-laminated timber (GLT) and Parallel Strand Lumber (PSL) columns of the Brock Commons were fabricated from locally sourced Douglas Fir trees. GLT members consist of individual pieces of dimension lumber that are bonded together with structural adhesives, creating a single element with highly rated strength properties [9]. Due to the variation in structural strength and the wide

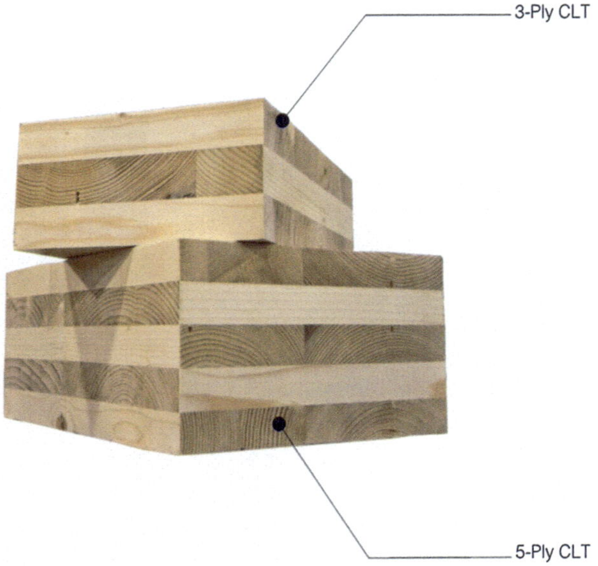

Fig. 10.18 Example of CLT panels formed from layered dimension lumber

range of available species and grades, glulam components are a common choice for columns and beams in mass timber structures. For this project PSL columns were used in high stress areas around the cores on the lower levels while GLT columns were used everywhere else. In addition to the columns, Cross-Laminated Timber (CLT) was used for floor panels, ultimately eliminating the need for load-carrying beams. Not unlike flat plate concrete construction, the two-way spanning capacity of CLT enabled the elimination of beams in favor of direct point load support on the columns. This permitted the transfer of axial loads directly from column to column, thus circumventing any crushing of the CLT panels. The lumber used to create the CLT panels was sourced from three separate companies located in British Columbia where the structure was built. CLT also consists of layers of dimensional lumber joined together with structural glue. However, the laminations are typically oriented at right angles to one another and glued to form larger panels of custom dimensions (Fig. 10.18). As a particularly cost-effective timber element, CLT is most commonly used as panel applications as floors, walls, and roofs (Fig. 10.19).

Timber is the primary structural element, amounting to a total volume of 78,857 cubic feet of wood for both the glulam members and the CLT panels [36]. Other than timber components, it is necessary to touch on the elements that allowed for timber to be used structurally in the upper floors of the building. The concrete podium at the ground floor in addition to the two concrete cores, spanning the height of the entire structure, was completed in the first phase of construction, lasting 7 months (Fig. 10.20). Although cast-in-place concrete added a considerable amount of time to the construction schedule, it is a strong material that provides lateral stiffness and stability, reduces cost, and ultimately simplifies the permitting process.

Fig. 10.19 Brock Commons. Assembly detail. GLT Columns and CLT Floors

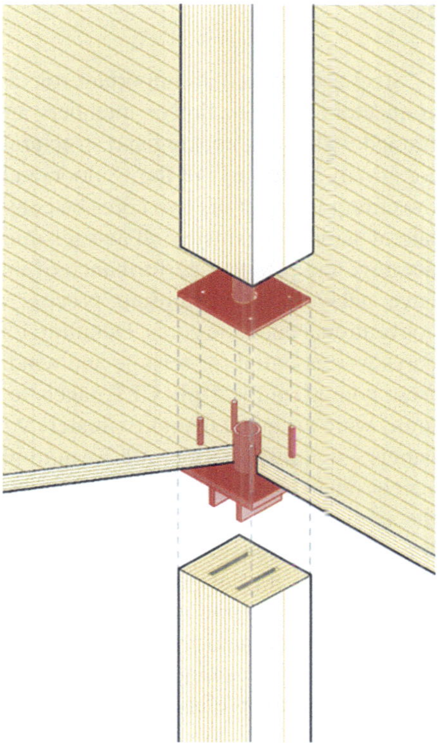

Fig. 10.20 Image during construction with visible concrete core and ground floor podium. (Image Credit: naturallywood.com. Photographer: KK Law)

Steel also played an important role as a structural material, both when used as connectors between timber members and in the framing of the roof. The roof is framed with metal decking and structural steel beams. Mass Timber is the structure for gravity loads in 17 of the 18 floors in Brock Commons. The height of this timber tower required that the two cores be constructed of concrete to provide the requisite lateral stability against wind and seismic loads. Additional time for research, engineering, and proof of performance of untested cores constructed with CLT would have delayed the permitting process and were therefore evaded in favor of conventional concrete cores. Detailing for ductility in the concrete shear walls of the cores was essential for seismic and wind forces. In addition, concrete topping slabs were poured on the 5-ply CLT panels to reduce wind-induced vibrations. The columns were arranged in approximately 9 ft. by 13 ft. to optimize the size of the CLT panels for transportation and to minimize waste. Brock commons is fitted with accelerometers, inclination gauges, and string potentiometers to constantly measure the axial shortening of columns, while moisture meters were also installed to continuously keep records of moisture content of the timber.

The decision to use mass timber in Brock Commons required a more intensive 3D modeling integration and prefabrication stage in contrast to concrete or steel buildings of the same size. All CLT and glulam components were coordinated and milled via CNC machines prior to being trucked to the site for erection and assembly. Although the prefabrication and coordination period required more time (about 3 months) than a comparable building, the erection and assembly time was exceptionally quick at 70 days (about 4 months faster than a typical project of the same size constructed in a material other than mass timber) (Fig. 10.21) [36]. Combining both the fabrication and assembly time of this timber frame, and comparing that to the length of time needed to construct only the concrete elements in the building, it becomes clear that, despite the additional fabrication time, timber construction involves a shorter construction schedule than cast-in-place concrete. Not only is timber a more sustainable material choice than concrete because it stores, rather than emits, carbon dioxide, the quicker assembly time can have several positive effects on the environment and surrounding community. In addition, the selection of mass timber entails a significantly lower embodied energy and much less air and water pollution. The volume of wood used in the Brock Commons has a potential carbon benefit of 2432 metric tons of sequestered carbon dioxide. This volume of wood circumvents 679 metric tons of greenhouse gas emissions in the form of carbon dioxide. This is equivalent to 511 cars off the road for a year or the equivalent energy used to operate a home for 222 years [36].

Challenging Future Concerns

Possibly the greatest argument against mass timber buildings is the combustibility and fire resistance associated with wood. To address these arguments and ensure that the building would be safe and meet all fire protection codes, Brock Commons

Fig. 10.21 Construction process with visible 70% wood fiber panels on the facade during erection. (Image Credit: naturallywood.com. Photographer: KK Law)

required extensive research, oversight and demonstration of performance of fire protection compliance. The fire protection engineer assigned to this project, Andrew Harmsworth of GHL Consultants Ltd., even stated that about seventy percent of the pre-construction period was spent securing code approval for fire safety. The design underwent multiple peer review processes that involved panels of fire safety experts who recommended areas for further analysis and validated proposed safety measures. Larger pieces of wood (timbers) will inherently char and provide a certain level of fire protection. However, additional techniques were used to extend the intrinsic fire-resisting qualities of the wood used. Ultimately as a means of allaying fears of fire and expediting building permits, a schedule of complete encapsulation was implemented to ensure that all CLT and glulam components were surrounded by a minimum of three layers of TYPE X fire-rated gypsum board (Fig. 10.22), thus providing 1–2 h of fire protection for all structural members in this occupancy category. Timber chars at a rate of 1.5 in. per hour and creates a layer that slows down the combustion rate and allows safe evacuation of occupants. The full encapsulation of all mass timber components allowed the members to be smaller since the design did not depend on char protection. Another advantage of encapsulation was that members were concealed and thus could be less visually attractive and more economical. Other active strategies to combat the spread of fire include automatic sprinklers on all floors with a supply tank independent of the municipal water supply. There were also measures taken to prevent fires during the construction phase.

Fig. 10.22 Interior glulam columns before gypsum board was placed for fire protection. (Image Credit: naturallywood.com. Photographer: KK Law)

The architectural design of the building, involving small compartmentalized (dorm) rooms, also helps prevent the possible spread of fire or smoke in any portion of the structure. The final mass timber design achieved a level of safety that is equal to or better than a design utilizing alternate noncombustible elements.

The use of timber in large buildings comes along with other challenges that must be addressed, and that are not encountered in steel or concrete construction. The primary goal of column connections in this building is to circumvent the transfer of axial loads through the CLT floor panels. The solution included a connection technique that involved the fastening of base plates to the glulam columns with epoxy anchors. This technique, progressive collapse, blocks the spread of damage from a specific point so that it will not affect the building's structural stability [35]. This connection method technique additionally works to avoid any shrinkage of the panels from affecting the columns. Wind-induced vibrations also needed to be addressed as required by the NBCC 2015. A dynamic wind load analysis was required and a reasonable value for damping of 1.5% was estimated based on the building's height and program [35]. Building sensors continue to monitor damping values well beyond the structure's completion.

Case Study Conclusion

Brock Commons Tallwood House Student Residence completely dismantles the argument that timber construction is not suited for high-rise buildings. It serves as a reminder that not only can buildings over six stories be constructed of timber but it can be done in a quick, effective, and safe manner. Since its completion in 2017, Brock Commons has received recognition from 15 different committees, mostly concerning the sustainable design choices and structural innovation implemented throughout the design and construction phases [37]. Brock Commons is currently the tallest mass timber building in North America and has demonstrated that local timber is an economically viable and sustainable option for high-rise structures.

Carbon Footprint Graphic

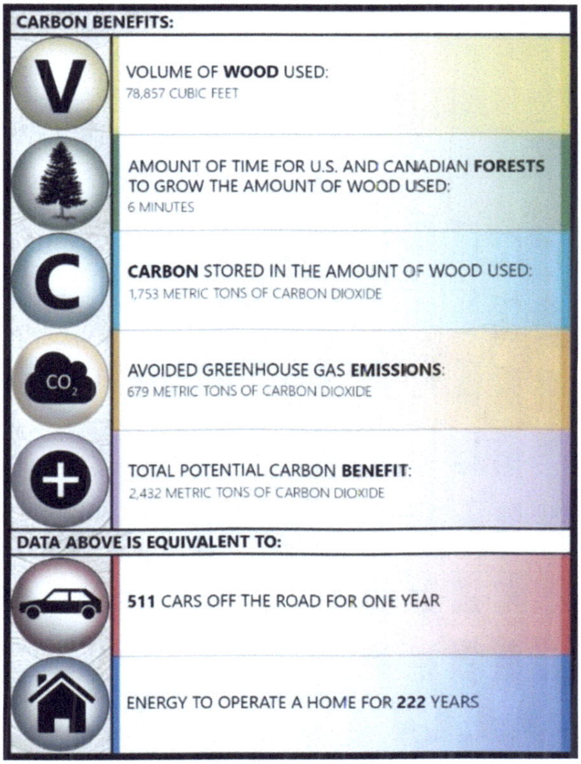

Case Study: T3 Minneapolis. Minneapolis, Minnesota

Architect: Michael Green Architecture (Fig. 10.23).

Introduction

Designed by Michael Green Architecture, T3 Minneapolis held the title of the largest mass timber office building in the USA at the time of its completion in 2016. The structure is also the first commercial building to be built of timber in the last 100 years. As Mr. Green alluded in his TED talk of 2013: "The nineteenth century was the age of steel, the twentieth of concrete, and this century promises to be of timber." This may prove to be true because climate change is turning the tide on concrete and steel construction in favor of more renewable and sustainable timber construction (Fig. 10.24).

Fig. 10.23 Timber. Technology. Transportation: T3 Minneapolis, MN. (Image Credit: James Peacock)

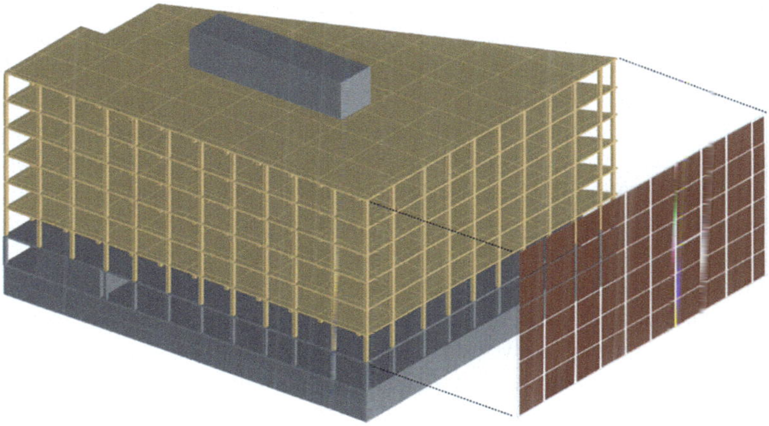

Fig. 10.24 T3 Minneapolis. Schematic Axonometric

With a height of seven stories and an area of 220,000 sq. ft., the timber assembly of T3 office building took only 9 weeks to erect, averaging about 30,000 sq. ft. per week [38]. With the goal of incorporating *T*ransit, *T*imber, and *T*echnology into the building, T3 structures are beginning to pop up around the US, Canada, and Australia. In addition to the T3 high end office building in Minneapolis, international real estate company Hines has developed other T3 buildings in cities such as Atlanta, Chicago, Denver, Melbourne, and Toronto. The inherent objective of these commercial real estate ventures is not only to use timber, a renewable material, but also to integrate sustainable transit objectives and innovative technological capabilities into a single structure. The design of T3 Minneapolis is inspired by its context of neighboring buildings in the warehouse historic district. It is conveniently located adjacent to a multimodal transportation hub and near light rail, commuter rail, bike paths, and other convenient transit nodes.

This building was the first WiredScore precertified building in Minneapolis. This certification rates the quality and resilience of the digital infrastructure of a building and ensures the best internet connectivity for users in all spaces including public spaces on the ground floor and the rooftop patio. The use of timber in office buildings has proven to be lucrative for tech-focused, coffee-sipping Millennial "hip" tenants, as evidenced by the fact that Amazon is the anchor tenant of this building. T3 buildings are on the forefront of commercial sustainable design and groundbreaking mass timber construction in North America.

Types of Wood Used

Like many other mass timber structures, the ground floor podium of the T3 Minneapolis is constructed of concrete, providing the foundation for the six floors of mass timber construction above it (Fig. 10.25). Unlike many of the other case studies in this chapter, the T3 project did not require special code approval. It follows the International Building Code (IBC)-TYPE IV-Heavy Timber construction, and thus does not exceed the allowable maximum height or area that is permitted by the prevailing code for this type of occupancy [39]. IBC TYPE IV construction does not specify a minimum fire rating but limits the height and area based on occupancy. In addition to the protective charring properties of wood and the fact that the building is sprinklered, an equivalent fire protection of 3 h is anticipated.

The majority of the timber used in this project came from trees destroyed by the pine mountain beetle and SFI-certified young trees, as opposed to full grown trees [39]. The choice to harvest these trees specifically illustrates the extent to which consideration was given in sourcing of the timber. One of the client's and architect's main goals was to have the least intrusive impact on the ecosystem. Compared to steel and concrete, timber construction requires a relatively small amount of energy to process raw logs into engineered wood products and the raw material is less

Fig. 10.25 Visible concrete podium and timber framing. (Image Credit: Blaine Brownell)

energy intensive to extract. According to the Green Building Council, 39% of CO_2 emissions in the USA are attributed to the production of cement and steel, not to mention emissions produced in transportation. Based on the US Forest Service, approximately half the weight of a tree is in the carbon it sequesters, making timber pound by pound stronger than steel. Buckminster Fuller believed that the strength-to-weight ratio of a material is an important measure of its structural efficacy.

Over 127,132 cubic feet of wood is used in the T3 office building with the majority being Nail Laminated Timber (NLT) [39]. NLT was chosen primarily for its structural advantages, low cost, and aesthetic quality. NLT consists of individual dimensional lumber components that are stacked on end and fastened on the wide face with nails or screws to create a single panel. Over 1100 NLT panels were assembled by StructureCraft for use in this project, totaling an area equivalent to that of nine hockey rinks (Fig. 10.26) [39]. All of the NLT in the ceilings, beams, and columns are left exposed on the interior of the project. In contrast, the NLT floors are capped with sound insulation material and a topping of concrete to create the desired acoustic effects and to reduce any vibration of the panels. In contrast to the encapsulated timber in Brock Commons, Michael Green Architects designed T3 with the intent of leaving as much timber as possible exposed to showcase its texture and warmth. This highlighted the desired aesthetic quality of the wood. Leaving the wood exposed by omitting interior finish materials reduces the possibility of having toxins and the cost of the structure.

Glulam is the timber selection chosen for the beams and columns used in T3 Minneapolis. In this project, glulam beams rest on notched columns that are

Fig. 10.26 One of the 1100 NLT panels assembled in T3. (Image Credit: Blaine Brownell)

reinforced with steel plates and fitted with ledger steel angles. Glulam columns support glulam beams spanning in the 25′ direction, and NLT panels spanning across the shorter 20′ direction make up the uniform structural grid. The NLT panels are 8 ft. wide and 40 ft. long and make up a continuous 2-bay span.

Leaving the columns and beams exposed not only limits the need for interior finishes but also creates a modern warm aesthetic that was desired for this urban office space. It is estimated that the T3's timber construction is 60% lighter than a comparable building made of concrete and 30% lighter than a similar structure built with structural steel [39]. In addition to the timber structure, the exterior of the building is designed to incorporate the character of the Minneapolis warehouse district with a progressive twist. The exterior envelope is fabricated from pre-weathered Corten steel panels, aluminum panels, and board-formed concrete. The T3 building is designed to appeal to young professionals while retaining the rustic character of the exposed steel on the exterior and the warm exposed timber on the interior. The greatest advantage of a timber structure over a similar steel or concrete structure is that it is much lighter which allows smaller support members and a smaller foundation. Another advantage of having a lighter structure is that it is designed for smaller seismic base shear. All of this translates into a more affordable and yet more sustainable structure (Figs. 10.27 and 10.28).

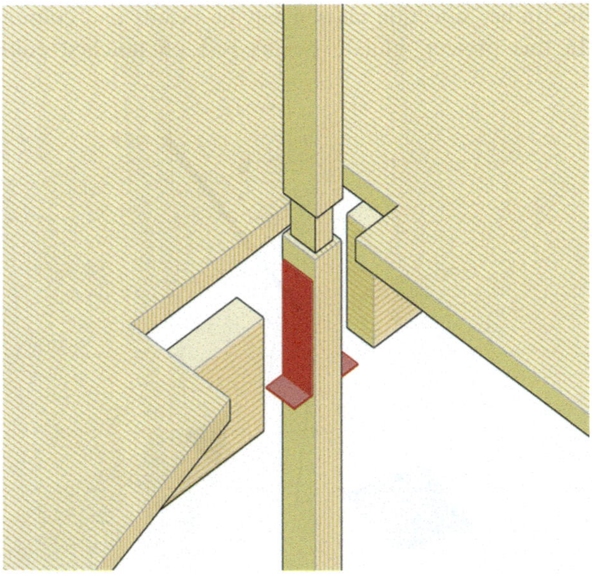

Fig. 10.27 Interior column/beams/NLT floor

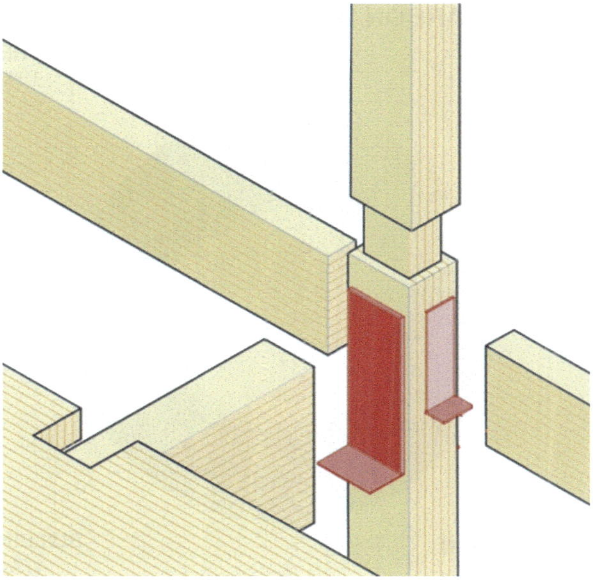

Fig. 10.28 Exterior column/beams/NLT floor

Environmental Sensitivity

The T3 Minneapolis office building is meant to integrate into its contextual environment yet offers high-tech amenities while utilizing energy-efficient systems. In a 2013 TED Talk about timber construction, Michael Green stated that nearly half of the world's greenhouse gas emissions are tied to buildings and up to a third are related to transportation (Fig. 10.29) [40]. Not only did Michael Green Architects attempt to lower greenhouse gas emissions by advocating for timber construction but also by encouraging other means of transportation to and from the construction site. Located in the North Loop Neighborhood of Minneapolis, the T3 site is commuter friendly and integrated alongside a bike trail. This location encourages the use of mass transit and alternate means of transportation. The ground level also includes a bike repair area and a bike storage space for tenants who choose to bike to the site instead of driving. In addition to bike repair and storage, the entire ground floor is reserved for retail space and a tenant amenity center. The T3 Minneapolis building avoids greenhouse gas emissions by an amount of 11,411 metric tons of carbon dioxide and has been awarded the certification of LEED Gold.

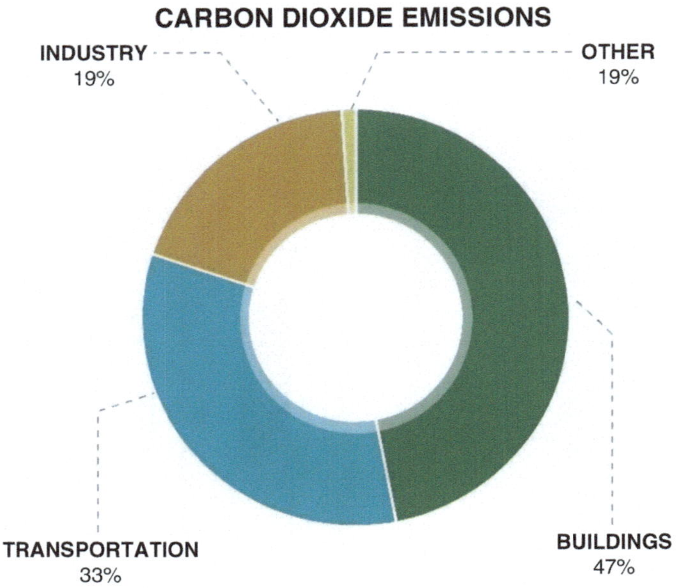

Fig. 10.29 Distribution of carbon dioxide emissions by field. (Based on Information from Architecture2030)

Case Study Conclusion

The T3 Minneapolis Office Building is proof that a modern timber construction building does not need to involve painstaking code exceptions or additional construction time. The office building is cost effective and was erected in an extraordinarily brief timespan. It followed all applicable code restrictions. Although timber construction is a primary element in this structure, T3 buildings focus on sustainability though transit and technology in addition to the use of wood. Since its completion in 2016, the T3 Minneapolis Office Building has received recognition from six different committees, primarily concerning the sustainable design choices and structural innovation implemented throughout the design and construction phases [41]. T3 Minneapolis showcases the possibilities of future office buildings by highlighting the practicality of sustainable design approaches.

Carbon Benefits Graphic

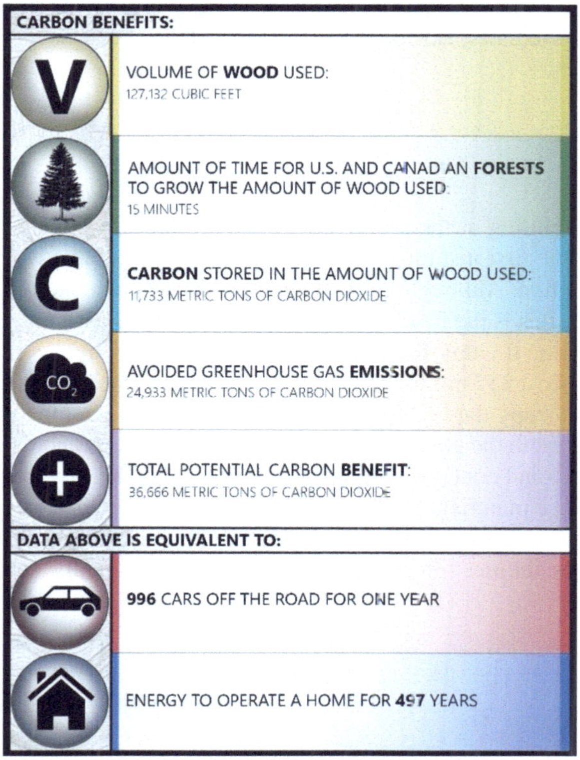

Case Study: John W. Olver Design Building at the University of Massachusetts Amherst

Architect: Leers Weinzapfel Associates, Boston, MA (Fig. 10.30).

Introduction

The John W. Olver Design Building at the University of Massachusetts Amherst (UMASS Amherst) is one of the first institutional buildings in the USA that advocates for mass timber construction. Although mass timber construction is common on the West Coast of the USA, this is the first building of its kind on the East Coast. With a total area of 87,600 sq. ft., this four-story structure designed by Leers Weinzapfel Associates (LWA) cost $52 million to build and serves as a teaching tool for the future designers who occupy the building [42]. Acting as a single homogeneous element in this building, both the glulam-steel hybrid "zipper" trusses and the Cross Laminated Timber (CLT) and concrete composite panels display the successful union of wood with another material. The unconventional cantilevered forms fabricated from wood featured in this project are examples of unprecedented techniques that illustrate the possibilities of timber construction, specifically the

Fig. 10.30 The John W. Olver Design Building. (Image Credit: Ajla Aksamija)

two-way spanning capabilities of CLT. Often regarded as one of the most advanced mass timber buildings in the world and the longest span in CLT, these hybridized members demonstrate the University of Massachusetts-Amherst's commitment to sustainability and innovation.

Hybrid Timber Elements

Possibly the most innovative and certainly the most eye-catching timber members in the John W. Olver Design Building are the glulam-steel zipper trusses. Positioned directly above the three- to four-story interior atrium, the seven trusses are most strikingly elegant and beautiful. The "New England" common space was the basis of design for the trapezoidal atrium that is 84 ft. long. The shorter end of the trapezoid is where the soil is weaker and the end truss is shortest at 31 ft. Where the soil is better on this sloping site, the zipper truss spans 55 ft. on the long end. The atrium houses tiered seating, a single-run CLT stair, and a ramp. The truss system was designed to accommodate skylights and the heavy loads of snow and a rooftop garden above. The garden is designed to include extensive and intensive green roof areas that serve as an assembly space and outdoor classroom. The trusses are spaced 12 ft. on center and vary in depth from 7 to 9 ft. deep, with a single point of convergence on each of the seven trusses (Fig. 10.31) [42]. Glulam components are used

Fig. 10.31 Photo of the "zipper" truss system in the main atrium. (Image Credit: Ajla Aksamija)

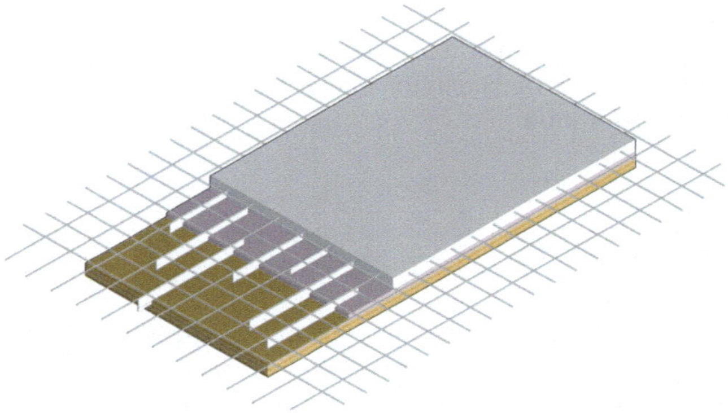

Fig. 10.32 4″ reinforced concrete topping on top of 1″ rigid insulation connected to 5-Ply CLT panel using steel composite connectors

for the compression members that all converge to a custom-designed steel plate on each truss. Each glulam element terminates with a custom cast steel "bullet" node connector. The tension members of each truss are configured as tension rods hanging between the bullet node and the end supports. This method of truss construction involves prefabrication of glulam elements and on-site assembly of the combined elements. The steel connectors were molded and cast by Cast Connex of Toronto. In addition to supporting the large assembly roof loads, these components highlight the cost-effectiveness of prefabricated timber elements and the dynamic forms that can be accomplished as a result of the hybridized truss systems.

Aside from the notable "zipper" truss system, mass timber is implemented in multiple other locations in the building. Glulam members are positioned as columns and beams with the intention of supporting the timber-concrete topping floor system on a 25 ft. × 25 ft. column grid. The composite floor panels are comprised of a 4-inch concrete topping slab site-cast on 5-layer CLT panels using patented shear connectors (Fig. 10.32). As a result, the concrete and CLT fuse and act as a single composite member with great benefits. The panels exhibit a high level of fire resistance, strength against bending forces, sound attenuation, and vibration control. Often with mass timber construction, vibration of the panels controls the design. Many of these unique material choices were the direct result of the original building design in steel and not in timber. When the choice to use timber was reached, many productive changes were implemented. With the use of the CLT-concrete assembly floor panels, the number of beams originally planned was reduced almost by 50% [43].

In addition to the timber used in the zipper trusses, other applications of mass timber in this building are in the use of CLT in the shear walls and glulam for diagonal bracing. The 60 ft. 7-layer CLT panels used as shear walls fully enclose and provide full fire protection to two stair shafts, the elevator shaft, and four service shafts in the building. One stair shaft and the elevator shaft had to be spliced because they exceeded both the manufacturer and transportation limitations. Each CLT shear wall panel was cut and milled using Computer Numerical Control (CNC) machines before arriving on-site, and subsequently assembled with screws and force-transfer connectors. To comply with the fire resistance requirements of a four-story shaft, each CLT shaft wall is covered with two layers of gypsum board and one layer of 1-inch noncombustible and moisture-resistant shaftliner panel on the exterior side only. This allowed the CLT panels to remain exposed on the interior of the shaft walls.

In accordance with the CLT walls and composite floor panels, the roof system is additionally constructed of 5-layer CLT panels. In total, the volume of wood used in the John W. Olver Design Building amounts to 72,467 cubic feet. This volume of wood can be grown in US and Canadian forests in only 6 min and has a potential carbon benefit of up to 2532 metric tons of carbon dioxide. This is equivalent to 535 cars off the road for 1 year or the amount of energy used to operate 267 homes for 1 year [42].

Designed on the basis of the International Building Code (IBC) of 2012, the design of the Design Building had to file many variances because CLT was not recognized by the IBC until 2015, and as an "alternate material." There will be many changes coming in the IBC to accommodate mass timber innovations, but not until the 2021 code. Leaving the CLT shear wall exposed in the cores and adding two layers of gypsum board on the outside to secure a 2-h fire rating was a big challenge. There had to be two panels of experts; one to approve the soundness of the structural engineering and the other to certify the fire safety of the construction. At the time of design and construction the building code was performative instead of prescriptive as the IBC 2021 promises to be. More often than not, this is the process with novel materials and systems.

The John W. Olver Design Building emphasizes and teaches energy performance and sustainability through its architectural design. Although the building was initially awarded LEED Gold certification, recent calculations show that the building may achieve LEED Platinum status due to energy saving features. While the building was designed to meet the Massachusetts stretch energy code requirement of producing 20% more energy compared to the code-specified baseline, it is actually producing an improvement of 50% [42]. Aside from the use of timber construction, some key elements that contribute to this level of energy savings include the use of glazing and skylights to provide maximum daylighting, specifically in the main atrium (Fig. 10.33). Furthermore, in an attempt to secure LEED certification, the building underwent a Life Cycle Assessment (LCA) that yielded comparative results between the mass timber design and the previous design in structural steel. This project also demonstrated the cost-effectiveness of mass timber construction as it cost only 4% more than the same design in steel. Among other sustainability data, the LCA confirms that the timber construction on this project is less likely to deplete non-renewable energy resources than the steel design by an amount of 14.8% [42]. The LCA furthermore provides concrete data demonstrating the sustainable benefits of timber construction in this specific building. Other sustainable features of the building include walkable access to public transportation, diverting significant amounts of waste from the landfill, and mitigating the heat island effect with the green roof. Other features include

Fig. 10.33 Photo of the main atrium with the "zipper" truss system, skylights, CLT staircase, and tiered seating. (Image Credit: Ajla Aksamija)

FSC-certified wood, radiant floors, heat recovery systems, and motion sensors among several others that earned it the high LEED certification.

Comparative Benefits of CLT

If designed in concrete, the weight of this building would be nearly six times that of timber. To resist seismic forces associated with the additional weight of an equivalent concrete structure, all elements and foundations would also have to be designed to a much more stringent standard. Seismic base shear is directly proportional to the dead load of the structure. As a result of the final lighter weight of the structure, timber is a more ductile and much lighter material choice to combat these dynamic forces (Fig. 10.34).

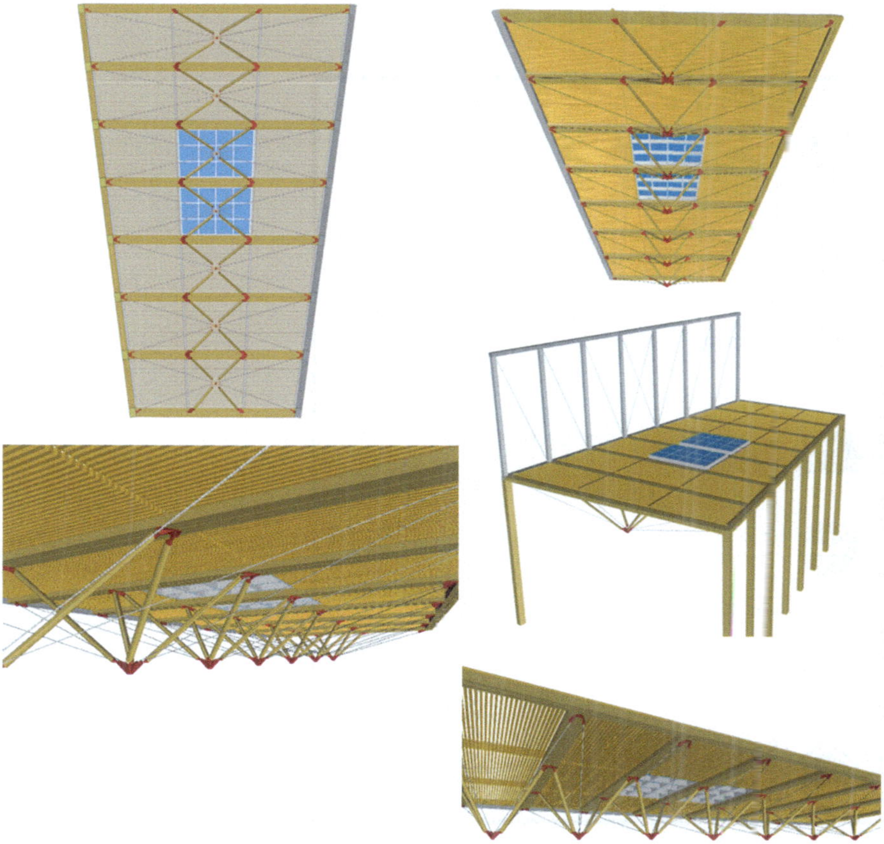

Fig. 10.34 Atrium reflected ceiling plan and views of the sipper trusses

Case Study Conclusion

The John W. Olver Design Building at the University of Massachusetts Amherst demonstrates the innovation and advanced capabilities of hybridized timber elements. The comparative data between the original steel design and the final mass timber design is evidence of the relevance of timber construction under certain circumstances. Since its completion in 2017, the John W. Olver Design Building has received recognition from 14 different committees, mostly concerning the sustainable design choices and structural innovation implemented throughout the design and construction phases [44]. Not only does this structure serve as an illustration of the creative possibilities of mass timber in contemporary architecture, it also houses future design professionals who are likely to learn design excellence through osmosis.

Carbon Benefits Graphic

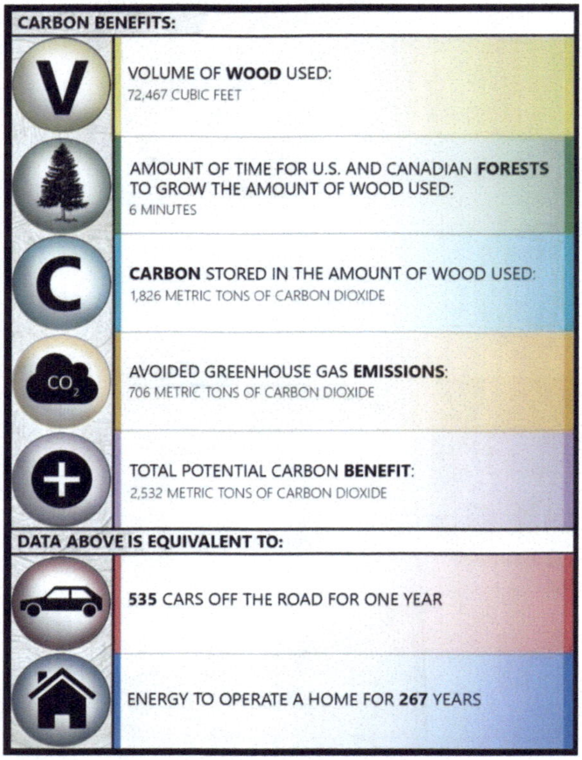

10.4 Conclusion

CLT is one of the most promising wood alternatives to concrete, particularly for non-residential and mid-rise markets around the world. Architects, engineers, and manufacturers have leveraged CLT and other Mass Timber products for their advantages, including design flexibility, aesthetics, strength to weight ratio, and overall material performance. There is a revolution in the building sector that is working to elevate timber to the level of steel and concrete, mostly through taller and longer spanning structures.

As mass timber and specifically CLT is relatively new to the construction market, it is hard to be confident about predictions that involve the recyclability of materials at the end-of-life stage of a building. The ideal option is to Design for Disassembly (DfD) so that panels may be salvaged during demolition and used elsewhere at a later date [19]. Another factor that is currently impeding the wide use of mass timber is that its cost per unit area is higher than that of steel or concrete. This cost concern may become less pronounced when consideration is given to savings gained in the speed of construction, minimizing site storage needs, smaller crane loads, and the reduction of construction waste. However, this amount is likely to be the same as the traditional methods if we consider reduced waste and reduced working hours that were mentioned earlier [14].

In addition to many aesthetic and structural performance opportunities described in this chapter, and as a renewable resource with low embodied energy, CLT offers a low carbon footprint while also sequestering and serving as a carbon sink. Interest in CLT as a new engineered wood product in North America is still in the early stages of development and is widely and rapidly proliferating. CLT provides an opportunity for the North American wood industry to build both larger and taller structures in wood. However, development is delayed by building regulations that cautiously safeguard the public health, safety, and welfare whenever a novel building product is introduced, materiality notwithstanding.

References

1. Forman, B. M. *Mill sawing in the seventeenth-century Massachusetts*. https://hne-rs.s3.amazonaws.com/filestore/1/2/8/0/0_743f3ba38c50e98/12800_43708f204bfc452.pdf
2. Schons, M. (2011). The Chicago fire of 1871 and the 'great rebuilding'. *National Geographic 25*. https://www.nationalgeographic.org/article/chicago-fire-1871-and-great-rebuilding/
3. "Review of Structural Materials and Methods for Home Building in the United States: 1900 to 2000." U.S. Department of Housing and Urban Development Office of Policy Development and Research. https://www.huduser.gov/portal/Publications/PDF/review.pdf
4. Williamson, T. G., & Borjen, Y. *Standard practice for the deprivation of design properties of structural glued laminated timber in the United States*. International Council for Research and Innovation in Building and Construction. https://www.apawood.org/Data/Sites/1/documents/technicalresearch/paper-2007-cib-w18%2D%2D-glulam-stresses.pdf
5. Schwarzmann, G. *Establishing new markets for CLT—Lessons learned*. Oregon State University.

6. *Looking back on 25 years of FSC*. Forest Stewardship Council. https://fsc.org/en/about-us/25-years-of-fsc
7. Evans, L. Cross laminated timber: Taking wood buildings to the next level. *reThink Wood*, American Wood Council, and FP Innovations.
8. Think Wood. (2016). *"Mass Timber in North America." Aia Course #K1609D* (pp. 1–12) [Online]. https://continuingeducation.bnpmedia.com/courses/think-wood/mass-timber-in-north-america/1/
9. Mass timber in North America, expanding the possibilities of wood building design. *Think Wood*.
10. Wood Works, Woods Products Council. (2018). Acoustics and mass timber: Room-to-room noise control. Wood Works.
11. The Beck Group. (2018). Mass timber market analysis. Council of Western State Foresters. https://www.oregon.gov/ODF/Documents/ForestBenefits/Beck-mass-timber-market-analysis-report.pdf
12. Martin, O. (2017). *Is mass timber really sustainable?* (pp. 1–5) [Online]. https://archpaper.com/2017/11/timber-construction-sustainable/
13. Darby, H. J., Elmualim, A., & Kelly, F. (2013). A case study to investigate the life cycle carbon emissions and carbon storage capacity of a cross laminated timber, multi-story residential building. In *Proceedings of the world sustainable building Conference* (pp. 1–8).
14. Timber, C. L. (2020). *What are the sustainability benefits of using cross laminated timber in construction?* (pp. 1–4).
15. Friendly, E. *The case for tall wood buildings how mass timber offers a safe, economical, and environmental friendly alternative for tall building structures* (2nd ed.). MGA, Michael Green Architecture.
16. Clt, W. H. Y., Us, A., & Us, C. "1800 88 72 44", 1800.
17. Pierobon, F., Huang, M., Simonen, K., & Ganguly, I. (2019). Environmental benefits of using hybrid CLT structure in midrise non-residential construction: An LCA based comparative case study in the U.S. Pacific northwest. *Journal of Building Engineering, 26*, 2–4. https://doi.org/10.1016/j.jobe.2019.100862
18. Guo, H., Liu, Y., Meng, Y., Huang, H., Sun, C., & Shao, Y. (2017). A comparison of the energy saving and carbon reduction performance between reinforced concrete and cross-laminated timber structures in residential buildings in the severe cold region of China. *Sustainability, 9*(8), 1426. https://doi.org/10.3390/su9081426
19. Astbury, J. Feature: Just how sustainable is cross-laminated timber? *Buildings: The Architects Journal*, 1–5. 201AD, [Online]. https://www.architectsjournal.co.uk/buildings/feature-just-how-sustainable-is-cross-laminated-timber/10024485.article
20. Hafner, A., & Schäfer, S. (2018). Environmental aspects of material efficiency versus carbon storage in timber buildings. *European Journal of Wood and Wood Products, 76*(3), 1045–1059. https://doi.org/10.1007/s00107-017-1273-9
21. Scarlet, T. (2015). *Cross laminated timber as sustainable construction technology for the future*. Helsinki Metropolia University of Applied Sciences. https://www.theseus.fi/bitstream/handle/10024/102020/Bachelor%20Thesis_Tommaso%20Scalet.pdf?sequence=1
22. Keene, S., & Smythe, C. (2009). *End-of-life options for construction and demolition Timber waste: A Christchurch case study*. University of Canterbury. http://citeseerx.ist.psu.edu/viewdoc/download?doi=10.1.1.629.5114&rep=rep1&type=pdf
23. Liu, Y., Guo, H., Cheng, S., & Chang, W. S. (2016). *Assessing cross laminated Timber (CLT) as an alternative material for mid-rise residential buildings in cold regions in China—A life-cycle assessment approach*. https://www.mdpi.com/2071-1050/8/10/1047/htm
24. Wilson, A. (2013). America's Greenest Office Building. Green Building Advisor. *Greenbuildingadvisor.com*, The Taunton Press, Inc.
25. Wood shines in sustainable 'Show & Tell.' Woodworks.org.
26. *Solar Panels*. Bulittcenter.org. https://bullittcenter.org/building/building-features/solar-district-1/

27. "Bullitt Center." Miller Hull Partnership, LLC. https://millerhull.com/project/bullitt-center/
28. Pena, R. "Living Proof: Seattle's Net Zero Energy Bullitt Center." University of Washington, Department of Architecture.
29. "*Regenerative Elevator*." Bulittcenter.org. https://bullittcenter.org/building/building-features/a-powerful-plunge/
30. "*Rainwater Harvesting*." Bulittcenter.org. https://bullittcenter.org/building/building-features/waterworks/
31. "*Greywater System*." Bulittcenter.org. https://bullittcenter.org/building/building-features/wastewater-use/
32. "*Radiant Heat*." Bulittcenter.org. https://bullittcenter.org/building/building-features/warmth-from-below/
33. "*Heat Recovery Ventilation*." Bulittcenter.org. https://bullittcenter.org/building/building-features/breath-of-warm-air/
34. *TallWood house at Brock Commons—Prefabricated timber at its peak*. The American Institute of Architects Continuing Education System.
35. The TallWood House at Brock Commons, Vancouver. *The Structural Engineer*. www.thestructuralengineer.org
36. "Brock Commons TallWood House." Naturally:wood. naturallywood.com
37. "Brock Commons Tallwood House." Acton Ostry Architects Inc. https://www.actonostry.ca/type/brock-commons/com
38. Wood: tenant-cool, tech-friendly commercial space. *Think Wood*.
39. Design and construction of taller wood buildings. *Think Wood*.
40. Callaghan, P. "Minneapolis' Office Building of the Future Will be Made of, Uh, Wood?" Minnpost. https://www.minnpost.com/politics-policy/2015/02/minneapolis-office-building-future-will-be-made-uh-wood/
41. "T3 Minneapolis." Michael Green Architecture. http://mg-architecture.ca/work/t3-minneapolis/
42. Inspiration through innovation at UMass Amherst, an exposed mass timber structure is a teaching tool. *Woodworks*.
43. A Journal of Contemporary Wood Engineering. Wood Design Focus. Forest Products Society, 2019.
44. John W. Olver Design Building. LWA-Architects. https://www.lwa-architects.com/project/integrated-design-building/
45. Mohammad, M., Gagnon, S., Douglas, B. K., & Podesto, L. (2012). Introduction to cross laminated timber. *Wood Design Focus, 22*, 3–12.

Chapter 11
Time and Nature for Responsive Wood Architecture. Two Projects of Schools' Buildings for Temporary and Adaptive Solutions

Antonella Trombadore, Gisella Calcagno, and Juan Camilo Olano

11.1 Introduction

Historical application of wood for structures and effective solutions in architecture were developed according to local climate conditions and technological human capacity to manage the natural resources. Wood is the oldest building material used by people for their homes and buildings. For a long time, wood has been combined with stones and bricks, but with the advent of reinforced concrete, it has been almost forgotten. In recent years, however, the characteristics and qualities of this sustainable construction material have been rediscovered: from lightness, seismic resistance, thermal insulation, and sound absorption, to renewability, natural beauty, and adaptability to the environmental conditions in which it is inserted. For this reason, every famous architect has always tried to use timber structures and around the world, there are many good examples of bioclimatic architecture and best practices properly inserted into the climatic and natural local conditions.

In this contribution two projects are analyzed to investigate innovative wood applications as dry construction systems for structural and envelope solutions,

The architectural design of *The Green Schools* in Selargius was elaborated by Prof Antonella Trombadore and Arch Juan Camilo Olano and submitted for the national competition for *Innovative Schools* promoted by MIUR (Ministry of Instruction, University and Research).
The design concept of the temporary wood pavilion was developed by the team of SPACE *BEX*Lab *Real and Virtual Space for Building Environmental Experience* of the DIDA Department of Architecture—University of Florence, in synergy with Municipality of Prato. The interdisciplinary working group, headed by prof. Saverio Mecca and Antonella Trombadore, was composed of prof. Flaviano Maria Giuseppe Lorusso (Design Concept), Prof. Matteo Zambelli, Arch. Gisella Calcagno (Integration of Technological solutions), and Arch. Juan Camilo Olano as BIM manager.

A. Trombadore (✉) · G. Calcagno · J. C. Olano
DIDA Department of Architecture, University of Florence, Florence, Italy
e-mail: antonella.trombadore@unifi.it

A. Sayigh (ed.), *The Importance of Wood and Timber in Sustainable Buildings*,
Innovative Renewable Energy, https://doi.org/10.1007/978-3-030-71700-1_11

implementing the aesthetical value of internal and external spaces, the energy performance, and the environmental quality of the "green buildings".

If we are used to thinking of architecture as a discipline that produces solid and permanent structures or to associate massive and heavy materials with the idea of building, we will be surprised to imagine what can be done by experimenting with alternative solutions with a material such as wood, proposing a renewed flexibility and a higher quality, also in this case of temporary structures.

Considering *time* as the fourth dimension of the project is an essential value not only in the approach to bioclimatic and adaptive architecture but also in the innovation dynamics of the urban structure, its cultural relations with the territory, and its possible regeneration processes. The concepts of temporariness and adaptability of architecture take on a completely contemporary character. In fact, if it is known from experience that temporary interventions often negatively affect the functional, morphological, environmental, and landscape quality of the contexts in which they are inserted, the design perspective must also concern the new meaning of temporality of architecture using time as the material itself of the project. This means designing new schools and educational spaces in terms of processes inherent to use/reuse/recycle, intervening on the existing structures through progressive and incremental grafting tactics.

11.2 The Concept of Green School: Wood as Environmental Friendly Material

We share the experience of a green school project in the municipality of Selargius (Sardinia), a little town in the Italian Sardinia island, as part of a national competition for *Innovative Schools* promoted by the Ministry of Instruction, University and Research (MIUR). Two ideas drive the concept design of the school as a creative ecosystem: the hexagonal cell-module, implementing the building identity as a vital organism; the grid of green infrastructure, that the building complex triggers on an urban scale, creating the green park to encourage an acceleration of environmental and social dynamics for a new urban metabolism, fostering innovation, and social involvement (Figs. 11.1 and 11.2).

Designed to encourage biodiversity, offering children a new way to learn environmental protection, the school complex is articulated as a green playground park: a wood building realized with environmentally friendly materials, a fluid plan that combines internal and external spaces, recreation areas, and teaching activities. The building is designed to create the best conditions of well-being, energy efficiency, environmental sustainability, and special attention to children's safety. The high level of indoor comfort is favored by the use of wood dry structure and eco materials, the optimization of natural daylighting and bioclimatic greenhouse, as well as the interaction of green spaces, the botanical garden, the green roof, and green facades, will implement the passive-evaporative cooling effect and environmentally conscious behavior of children (Fig. 11.3).

Fig. 11.1 External view of the green school in Selargius and integration of green spaces with the external garden

The project area is set in a network of sport areas and is also adjacent to three equipped green areas. To provide a *continuum* of green areas strongly inserted in the urban context, the school structure born as an aggregation of functional units that include a botanical garden and meeting places that can be used by the whole community. The project idea is based on the need to offer an educational environment enriched with good practices, in which the children may have the opportunity to express their needs and personalities, and find stimulus and support in their search for adequate answers, in a space defined by the dimensions of play, creativity, and circulation of ideas, where learning and even living the school spaces (Fig. 11.4).

Particular attention is paid to the articulation of the school spaces, conceived as a flexible multifunctional environment adaptable to the different educational needs, capable of accommodating groups of children of different ages, abilities, and interests, and the nature of the materials, to offer multiple opportunities for sensory and motor experiences, in order to increase the natural curiosity of children. Designed to encourage biodiversity by creating a green corridor between the city and the surrounding area, the new nursery school develops like a green playground park and offers children a new way to learn about environmental protection. It is a building complex made entirely of wood and environmentally friendly materials, with an articulation which blends indoor and outdoor spaces, recreation areas and teaching environments (Fig. 11.5). Everything is immersed in a large open-air laboratory in which to learn respect for nature and love for plants and animals.

The Wood Structure of Hexagonal Module as Cell of the Building Organism

There are two conceptual assets on which the design idea of the school as a creative ecosystem is based: building as a vital organism that leads to the architectural solution of the hexagonal cell-module; the network of nodes and connections that the

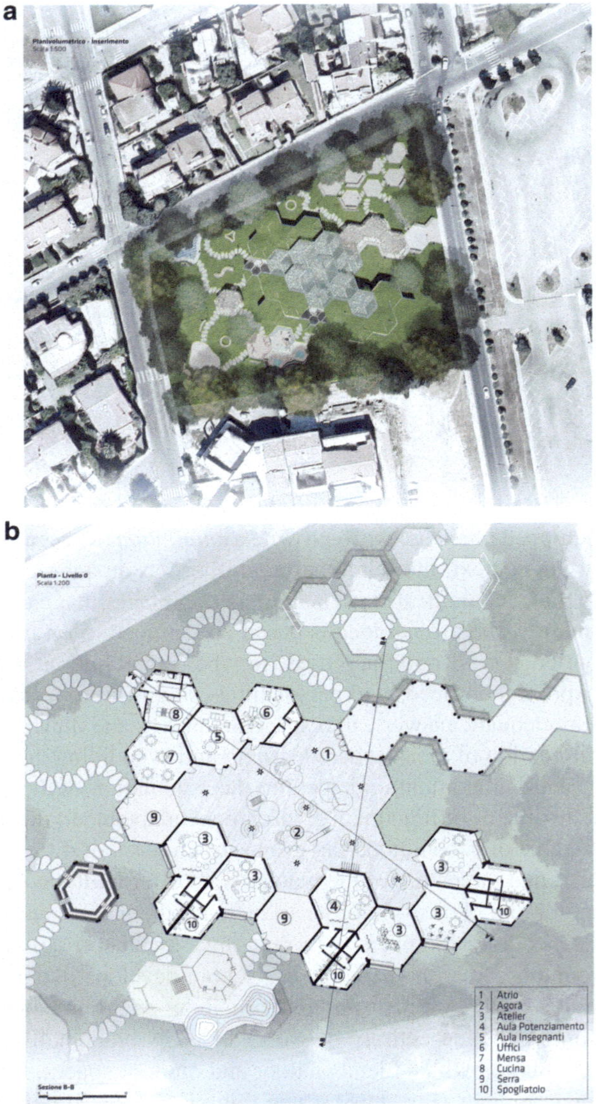

Fig. 11.2 (**a–b**) Plan of the green school and distribution of functions: Group spaces I Individual spaces I Spaces for exploration I Informal spaces I Agora

building complex triggers at the urban scale. The creation of the park and green infrastructures aims to promote an acceleration of environmental and social dynamics for a new urban metabolism: *innovation* and *participation*.

The elementary module, the hexagon, is the archetype of the cellular structure, easily aggregable, divisible, and distributable in space, also offering a multiplicity of combinations and connections. The sizing of the basic module of the hexagon ensures the surface area in accordance with legislation for the atelier/classroom. The shape also guarantees the subdivision and organization of the atelier/classroom according to the activities to be carried out.

Fig. 11.3 External view of the classroom and their relation with garden, designed as didactic space

Fig. 11.4 View of the interactive didactic space: Spaces for exploration, green space, and Agora

Fig. 11.5 View of the interactive didactic space: Spaces for exploration, green space, and Agora

The same module is used for the services connected with the school, atrium with changing rooms, canteen with kitchen and related services, administration, service personnel, and teachers' offices. Appropriately distributed in the aggregation module of hexagons, the project also provides for a bioclimatic Greenhouse. The play of the modules also includes their disintegration and dissolution in the progressive removal from the school structure, identifying Shaded spaces (gazebos), Recreation spaces with colored flooring and materials of various granulometry, garden spaces, botanical garden, and a body of water (Figs. 11.6, 11.7 and 11.8).

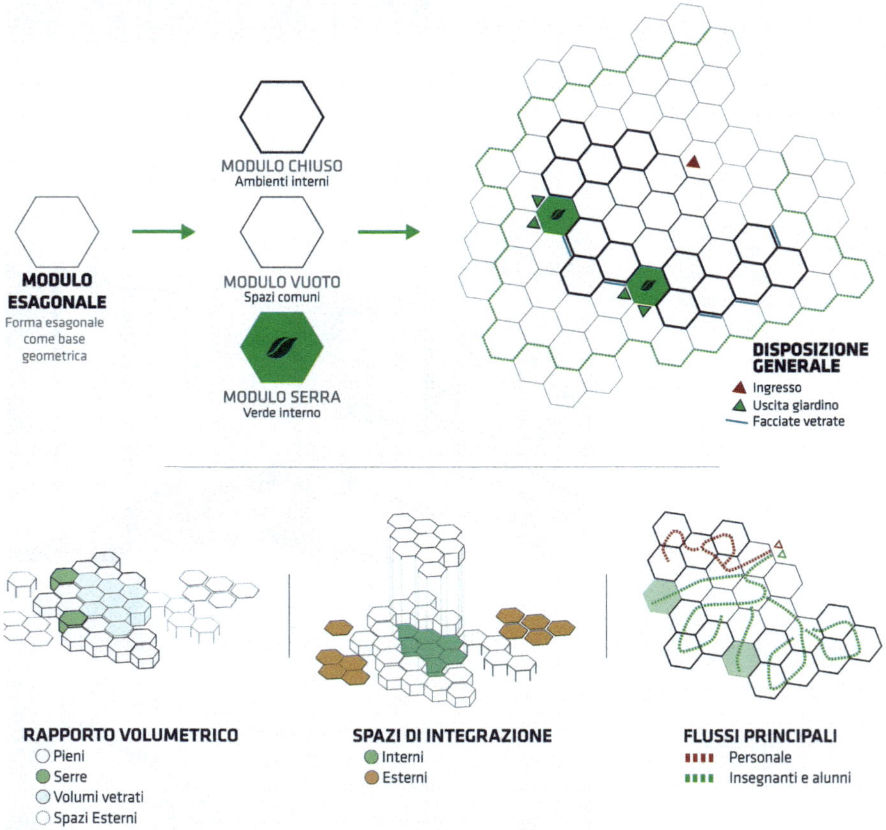

Fig. 11.6 Elaboration of the different scenarios of the plan. *The base module:* allows aggregation of one or more modules on six sides and its subdivision with fixed or mobile elements of standard size

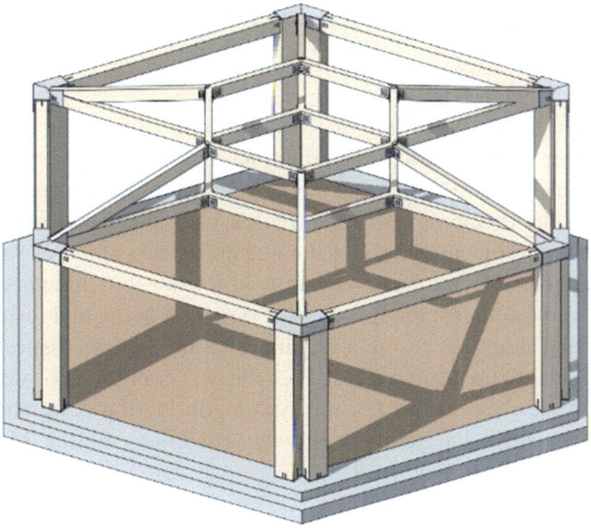

Fig. 11.7 The 3D representation of module where is possible to analyze the wood elements and structure

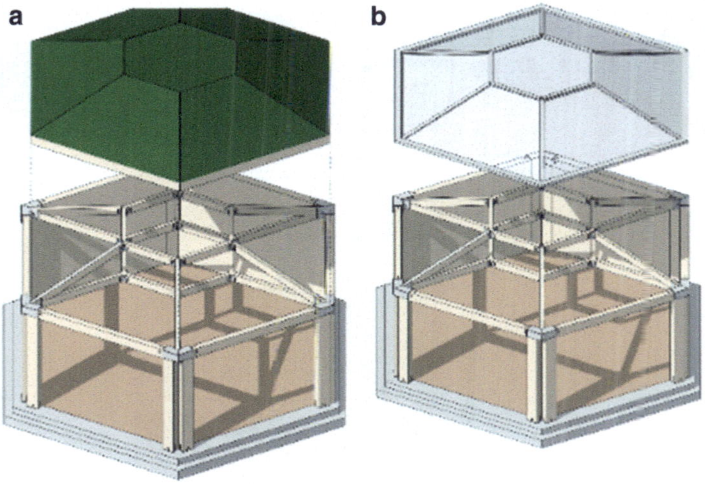

Fig. 11.8 (**a**, **b**) Two examples of roof solutions: on the left the integration of extensive green roof and on the right the opaque roof in which is possible to integrate PV systems

The Application of Wood as Eco-Logical Material for Resilience and Indoor Comfort

Choosing wood as the only building material, declining it in its formal and aesthetic potential, seeking architectural, construction, and technological solutions that would also enhance its textures, was the driver of the design approach, strongly oriented towards the theme of resilience. Increasing the effectiveness in the use of resources and the ecological quality of the technological capital of wood has made it possible to focus on the one hand on energy efficiency, bioclimatics, and the integration of renewable sources; on the other hand to well-being and respect for safety, without neglecting attention to the structural and anti-seismic aspect. The school building is designed to create and experience the best conditions of visual, acoustic, and thermohygrometric comfort. For this reason, during the design and simulation of energy performance, the specific climatic conditions of the Municipality of Selargius were considered (Fig. 11.9).

Fig. 11.9 (**a–c**) 3D view of didactic space. The quality of daylight is guaranteed by the architectural integration of skylight and light shelf

Fig. 11.10 (a, b) Integration of PV with double glass panel to allow the passive thermal contribution during the winter season

Energy and Environmental Well-Being and Sustainability

Indoor comfort is favored by the use of environmentally friendly materials (wood), by the optimization of natural lighting levels, by the presence of bioclimatic greenhouses, by heating strategies (direct solar gains) and passive cooling (solar shading and natural ventilation). The choice of orientation, the planimetric development, and the presence of the green roof, the garden, and the water that increase evaporative cooling are relevant for the local microclimate. The technological solutions provided and the integration of photovoltaics allow energy savings of 50% compared to classic school facilities, without increasing construction costs (Fig. 11.10).

Safety and Attention to Seismic Risk

The wooden construction, with its elasticity capacity, is well suited to be applied in seismic areas, to respond to both tensile and compressive forces due to static loads. Its lightness also limits the effects of the earthquake itself with the same magnitude and characteristics of the building, avoiding the collapse of the structure. The choice of wood for the load-bearing structure elements, the external envelope, and the internal partitions has further advantages such as thermal insulation, the breathability of the external envelope, and the presence of a natural insulation that reduces summer overheating. Furthermore, the hexagonal module, to be made dry with self-supporting partitions and a double ring cover that also contains a skylight in the drum, allows to significantly reduce costs and production times, guaranteeing maximum control over the quality and performance of the envelope and greater sustainability during construction process and site management.

The Positive LCA

By carrying out an analysis on the life cycle of the building and comparing it with buildings of the same size and made of concrete and other materials, starting from the production of the raw material, to the transport and construction phases, up to the final realization, it is clear that the construction in wood takes half the energy necessary to build the same building in reinforced concrete.

The life cycle assessment (LCA) allows to detect advantages in terms of time (prefabrication), lower quantity of waste to be disposed of, reduction of realization times, economic savings, reduction of energy consumption, lower environmental impact, and any recovery for recycling (Figs. 11.11 and 11.12).

Fig. 11.11 (**a–c**) 3D views of the green school and the integration with the greenspace, play area and the gardens

Fig. 11.12 External view of the green school and integration with the landscape

11.3 The Concept of Temporary Wood Pavilion: Multifunctional Expansion of External School Space

The project meets the need, expressed by the municipality of Prato, to create temporary structures to expand the educational spaces of schools in this period of COVID-19 pandemic, requiring a greater physical distance between students, by addressing the theme of outdoor didactics, based on a direct contact with nature and green areas. The temporary structures are for the primary schools (students aged 6–11) and the first years of secondary school levels (aged 12–15) that need to recover greater interaction with outdoor spaces, allowing them to carry out educational and recreational activities oriented towards a direct knowledge of natural, biological, and physical phenomena (for example how a plant grows, how algae develop, how fish live in an aquatic habitat, etc.). The idea is to create real educational workshops integrated in the context outside the school, full of natural and stimulating elements for the creative learning process of children.

The design idea was developed starting from a simple and clear concept: an open, covered volume, defined by the combination of two structural platforms, base and roof, delimited by a perimeter of vertical elements-pillars that mark a portico on two sides; a further extension of the platform performs a connective function with the existing building and with the green areas, becoming a distributive element that favors the circulation of students. The free volume identified in plan by the pillars represents a new model of outdoor/indoor educational space, in which it is possible to flexibly configure diversified educational spaces based on the innovative educational activities to be carried out, such as *Group spaces* | *Individual spaces* | *Spaces for exploration* | *Informal spaces* | *Agora* (INDIRE, Model 1+4 educational spaces).

The conceived wooden structure addresses the required flexibility for the overlapping functions and/or contemporaneity of teaching activities, as well as for the eco-compatibility of the material and for its aesthetic quality, particularly suitable for an architecture aimed at children and young people.

The choice of a dry construction in wood has considerable advantages not only in the immediate case of the COVID-19 emergency, requiring fast solutions, but also in the life cycle of the school building. The prefabrication of wooden elements and components, as well as their dry assembly, highly reduces the construction time, guaranteeing an easy management of the construction site and of the operational phase, such as the possibility of disassembly and reuse (Fig. 11.13).

Technical and Functional Characteristics

The temporary wooden structure develops with the following characteristics (Figs. 11.14 and 11.15 and 11.16):

Fig. 11.13 External view of the wood pavilion for the multifunctional expansion of the school space

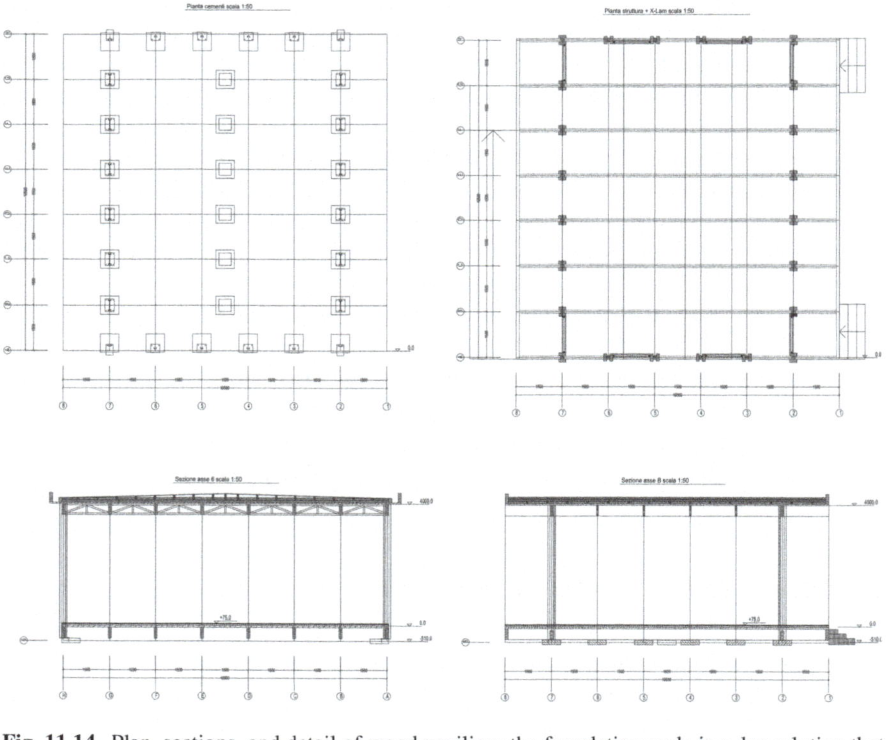

Fig. 11.14 Plan, sections, and detail of wood pavilion: the foundation node is a dry solution that will be linked to the concrete platform or metal screws

Fig. 11.15 (**a–c**) External and internal view of wood pavilion. The flexibility of the space allows to combine different activities

Fig. 11.16 (**a**, **b**) Internal and external view of the wood pavilion with the integration of movable shading devices and curtain

- Double platform made with double weaving of wooden reticular beam (10.5 m × 10.5 m): a base one, slightly raised from the ground (about 45 cm) and an upper one, perfectly symmetrical and overturned with respect to the basement, which is configured as a reticular structure to support the roof covering. The volume between the two plates is left open laterally, marked by vertical elements, double pillars in laminated wood. The sequence of the pillars thus defines the internal surface (10.5 m × 7.5 m) dedicated to the playful, didactic, and educational space, leaving two lateral corridors to the outside that perform the function of distribution and connective system with the existing building and the garden.
- Double pillars in laminated wood, spaced apart with a center distance of 1.5 m and a free distance between elements of 1.25 m, compatible with the Euro Standard measurement. This technological solution allows for greater flexibility

of the structure over time. In fact, the pavilion can be implemented over time, both from a dimensional and a performance point of view, satisfying the different needs for carrying out educational activities and indoor comfort in the different seasons. Such a flexible solution offers the possibility of enriching and shaping the envelope over time with light, transparent, mobile elements or with prefabricated brushes anchored directly to the frame of the pillars which, with different layering of materials, can give life to new architectural configurations and aesthetics but above all to reach new performance targets.
- Dry foundations of the screw type and light wooden structure completely removable and reversible in order to restore the area to its current state.

Modularity and Flexibility

The structure is set up on a modular geometric scheme, with a "basic" structural element consisting of a beam/pillar that is repeated with a center distance of 1.5 m. The dimension of the center distance corresponds to the Euro Standard measure of the prefabricated and pre-assembled panels on the market, in order to allow the installation of opaque and/or transparent curtain walls and windows, these too easily removable, to increase the level of flexibility and adaptability of spaces and the different possibilities of use of the structure at different times of the day and in different seasons.

Based on the needs of the client, as well as the outdoor spaces available in schools, it is possible to diversify the quantity of structural modules to be installed.

The following is a configuration defined to fit into two school contexts in the municipality of Prato: the garden, the Cesare Guasti elementary school, and the green space of the Filippino Lippi school. In this case, the structure is made up of eight modules, for a total length of 10.5 m. In addition to the supporting structure in laminated wood, also the infill and finishing materials have been chosen for their eco-compatibility:

- The foundations are made of steel on metal screws
- The raised load-bearing structural elements in laminated wood protected from soil moisture
- Tnsulation with natural fibers
- The roof covering is made of metal panels, with internal insulation in natural fibers

The temporary structure aims at energy saving: thanks to the integration of lighting bodies with LED technology, with manual and automatic controls and an electrical system powered by PV panels, the energy needs are met by the contribution of renewable energy, allowing autonomous energy (stand-alone) of the structure. In addition, particular attention is given to the use and accessibility of the structure for people with reduced mobility, through the integration of ramps or platforms that connect the pavilion with the main body.

11.4 Considerations and Results: Next Challenge and Future Trend

The results of the two projects indicate how it is possible to perform different functions, maintaining spatial flexibility and remaining faithful to the structural configuration, also in the case of a temporary and provisional solution, increasing the ecosystemic qualities of the spaces dedicated to the education of our children (Fig. 11.17).

Various scenarios of spatial configuration of the didactic structures have been formulated and shared with administrations, stakeholders (students and teachers), and wood companies, following collaborative design dynamics, activating a process to revitalize the school spaces (both indoor and outdoor), strengthening the identity of the places dedicated to teaching and the sense of belonging of the students, triggering new social and environmental balances. What elaborated is an architectural response that is no longer static but flexible and changeable, to transform the space into a system capable of reacting to the stimuli and needs of its ever-changing school

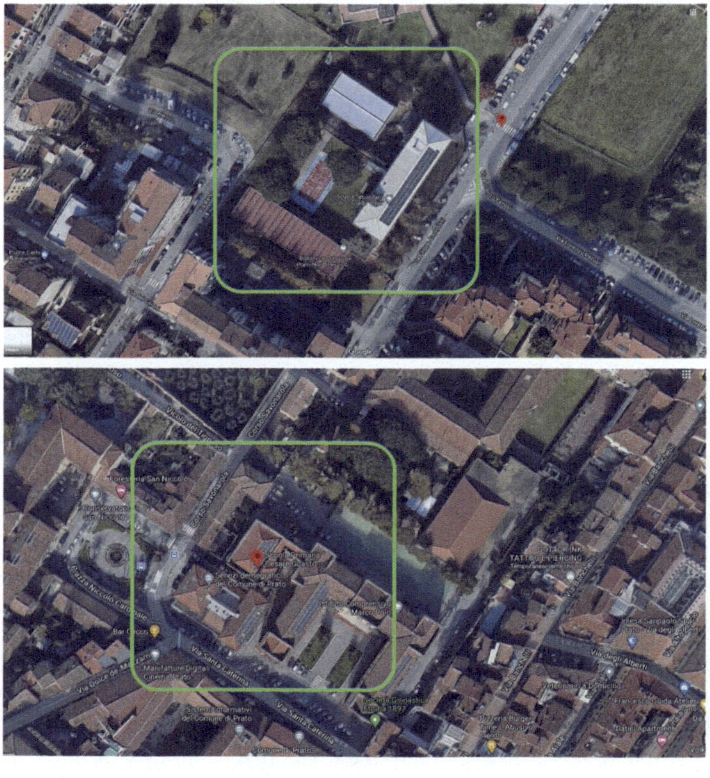

Fig. 11.17 Two different sites in the municipality of Prato where testing the solution: the secondary school Filippino Lippi and primary school Cesare Guasti

community. The research results have a high potential for replicability (on any area that presents a similar situation), highlighting the value of wood as a highly sustainable material (economic and technological) and the possibilities of using natural materials in the new typological reconfigurations, ensuring high levels of comfort, eco-compatibility, and energy efficiency.

Wood allows to understand the dynamics of architecture linked to the time variable, not only as a design exercise but also as the main asset to transit from a dissipative/linear economy to a regenerative/circular one for cities and sustainable urban contexts. Creative design explorations can configure capable architectures which can trigger new spatial dynamics and accelerate innovative educational and social contaminations. New scenarios for innovative educational spaces.

Bibliography

1. Antonini, E., & Boeri, A. (2011). *Progettare scuole sostenibili. Criteri, esempi e soluzioni per l'efficienza energetica e la qualità ambientale*. Palermo: Edicom Edizioni.
2. ARUP. (2014). *Cities alive—Rethinking Green infrastructures*. Foresight.
3. Natterer, J., Herzog, T., & Volz, M. (1998). *Atlante del legno*. Torino: Utet.
4. Andreu, D. (2019). *Legno. Architettura oggi*. Milan: Loft Media Publishing.
5. Trombadore, A., & Olano, J. C. (2018) Scuola come ecosistema creativo: l'esperienza progettuale per la green school a Selargius. In M. Fumo, G. Ausiello, & M. Buanne (Eds.), *Verso una Scuola Resiliente, Proceedings*. Napoli: Luciano Editore.

Index

A. Sayigh (ed.), *The Importance of Wood and Timber in Sustainable Buildings*, Innovative Renewable Energy, https://doi.org/10.1007/978-3-030-71700-1

The manufacturer's authorised representative in the EU is Springer Nature Customer Service Centre GmbH, Europaplatz 3, 69115 Heidelberg, Germany. If you have any concerns regarding our products, please contact ProductSafety@springernature.com

Printed and bound by CPI Group (UK) Ltd, Croydon, CR0 4YY
30/09/2026
02225695-0010